Modern Birkhäuser Classics

Many of the original research and survey monographs in pure and applied mathematics published by Birkhäuser in recent decades have been groundbreaking and have come to be regarded as foundational to the subject. Through the MBC Series, a select number of these modern classics, entirely uncorrected, are being re-released in paperback (and as eBooks) to ensure that these treasures remain accessible to new generations of students, scholars, and researchers.

Tata Lectures on Theta II

David Mumford
With the collaboration of
C. Musili, M. Nori, E. Previato,
M. Stillman, and H. Umemura

Reprint of the 1984 Edition

Birkhäuser
Boston • Basel • Berlin

David Mumford
Brown University
Division of Applied Mathematics
Providence, RI 02912
U.S.A.

Originally published as Volume 43 in the series *Progress in Mathematics*

Cover design by Alex Gerasev.

Mathematics Subject Classification (2000): 01-02, 01A60, 11-02, 14-02, 14K25, 30-02, 32-02, 33-02, 46-02, 55-02 (primary); 14H40, 14K30, 32G20, 33E05, 35Q99, 58J15, 58J60 (secondary)

Library of Congress Control Number: 2006936982

ISBN-10: 0-8176-4569-1 e-ISBN-10: 0-8176-4578-0
ISBN-13: 978-0-8176-4569-4 e-ISBN-13: 978-0-8176-4578-6

Printed on acid-free paper.

 Birkhäuser

9 8 7 6 5 4 3 2 1

www.birkhauser.com (IBT)

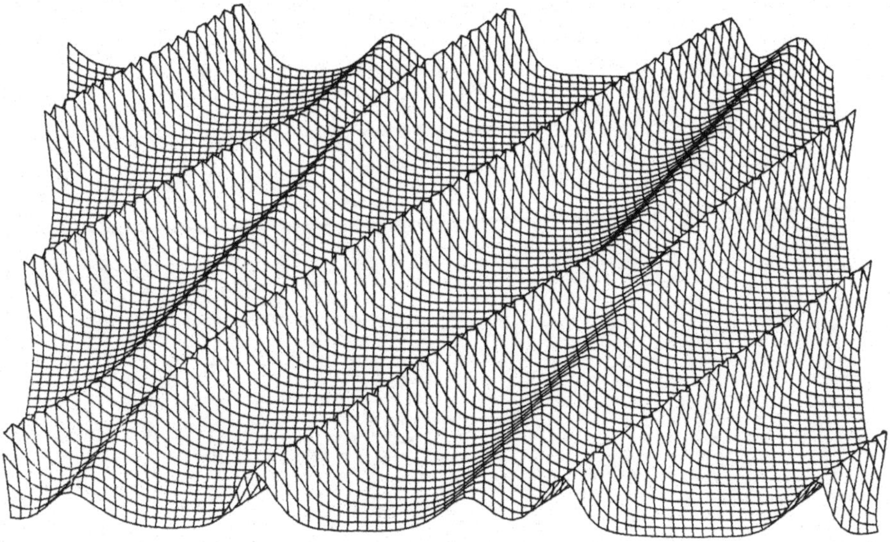

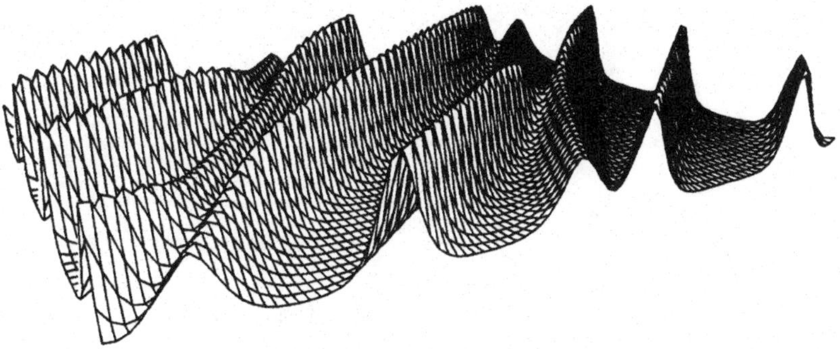

Almost periodic solution of K-dV given by the genus 2 $\mathfrak{p}$-function $D^2\log\text{-}\vartheta(z,\Omega)$ with $\Omega = \begin{pmatrix} 10 & 2 \\ 2 & 10 \end{pmatrix}$.

An infinite train of fast solitons crosses an infinite train of slower solitons (see Ch. IIIa, §10,IIIb, §4).

Two slow waves appear in the pictures: Note that each is shifted backward at every collision with a fast wave.

David Mumford

With the collaboration of

C. Musili, M. Nori, E. Previato,

M. Stillman, and H. Umemura

Tata Lectures on Theta II

Jacobian theta functions and differential equations

1993 Birkhäuser

Boston · Basel · Berlin

David Mumford
Department of Mathematics
Harvard University
Cambridge, MA 02138

Library of Congress Cataloging-in-Publication Data

Mumford, David.
Tata lectures on theta II.
Jacobian theta functions and
differential equations / with the collab. of C. Musili ... -
Boston; Basel; Berlin Birkhäuser, 1984.
Progress in mathematics; Vol. 43)
ISBN 0-8176-3110-0 (Boston)
ISBN 3-7643-3110-0 (Basel)

Printed on acid-free paper

Second Printing 1987
Third Printing 1993

ISBN 0-8176-3110-0
ISBN 3-7643-3110-0

Printed and bound by Edwards Brothers, Ann Arbor, MI
Printed in USA

9 8 7 6 5 4 3

CHAPTER III

Jacobian theta functions and Differential Equations

Introduction to Chapter III

In the first chapter of this book, we analyzed the classical analytic function

$$\vartheta(z,\tau) = \sum e^{\pi i n^2\tau + 2\pi i n z}$$

of 2 variables, explained its functional equations and their geometric significance and gave some idea of its arithmetic applications. In the second chapter, we indicated how ϑ generalizes when the scalar z is replaced by a vector variable $\vec{z} \in \mathbb{C}^g$ and the scalar τ by a g×g symmetric period matrix Ω. The geometry was more elaborate, and it led us to the concept of abelian varieties: complex tori embeddable in complex projective space. We also saw how these functions arise naturally if we start from a compact Riemann surface X of genus g and attempt to construct meromorphic functions on X by the same methods used when g = 1.

However, a very fundamental fact is that as soon as $g \geq 4$, the set of g×g symmetric matrices Ω which arise as period matrices of Riemann surfaces C depends on fewer parameters than g(g+1)/2, the number of variables in Ω. Therefore, one expects that the Ω's coming from Riemann surfaces C, and the corresponding tori X_Ω, also known as the Jacobian variety Jac(C) of C, will have <u>special properties</u>. Surprisingly, these special properties are rather subtle. I have given elsewhere (Curves and their Jacobians, Univ. of Mich. Press, 1975), a survey of some of these special properties. What I want to explain in this chapter are some of the special function-theoretic properties that ϑ possesses when Ω comes from a Riemann surface. One of the most striking properties is that from these special ϑ's one can produce solutions of many important non-linear partial differential equations that have arisen in applied mathematics. For an arbitrary Ω, general considerations of functional dependence say that $\vartheta(\vec{z},\Omega)$ must always satisfy many non-linear PDE's: but if $g \geq 4$, these equations are not known explicitly. Describing them is a very interesting problem. But in contrast when Ω comes from a Riemann surface, and especially when the Riemann surface is hyper-elliptic, ϑ satisfies quite simple non-linear PDE's of fairly low degree. The best known examples are the Korteweg-de Vries (or KdV) equation and the Sine-Gordan equation in the hyperelliptic case, and somewhat more complicated Kadomstev-Petriashvili (or KP) equation for general Riemann surfaces. We wish to explain these facts in this chapter.

The structure of the chapter was dictated by a second goal, however. As background, let me recall that for all $g \geq 2$, the natural projective embeddings of the general tori X_Ω lie in very high-dimensional projective space, e.g., $\mathbb{P}_{(3^g-1)}$ or $\mathbb{P}_{(4^g-1)}$ and their image in these projective spaces is given by an even larger set of polynomials equations derived from Riemann's theta relation. The complexity of this set of equations has long been a major obstacle in the theory of abelian varieties. It forced mathematicians, notably A. Weil, to develop the theory of these varieties purely abstractly without the possibility of motivating or illustrating results with explicit projective examples of dimension greater than 1. I was really delighted, therefore, when I found that J. Moser's use of hyperelliptic theta functions to solve certain non-linear ordinary differential equations leads directly to a very simple projective model of the corresponding tori X_Ω. It turned out that the ideas behind this model in fact go back to early work of Jacobi himself (Crelle, 32, 1846). It therefore seemed that these elementary models, and their applications to ODE's and PDE's are a very good introduction to the general algebro-geometric theory of abelian varieties, and this Chapter attempts to provide such an introduction.

In the same spirit, one can also use hyperelliptic theta functions to solve explicitly algebraic equations of arbitrary degree. It was shown by Hermite and Kronecker that algebraic equations of degree 5 can be solved by elliptic modular functions and elliptic integrals. H. Umemura, developing ideas of Jordan, has shown how a simple expression involving hyperelliptic theta functions and hyperelliptic integrals can be used to write down the roots of any algebraic equation. He has kindly written up his theory as a continuation of the exposition below.

The outline of the book is as follows. The first part deals entirely with hyperelliptic theta functions and hyperelliptic jacobians:

§0 reviews the basic definitions of algebraic geometry, making the book self-contained for analysts without geometric background.

§§1-4 present the basic projective model of hyperelliptic jacobians and Moser's use of this model to solve the Neumann system of ODE's.

§5 links the present theory with that of Ch. 2, §§2-3.

§§6-9 shows how this theory can be used to solve the problem of characterizing hyperelliptic period matrices Ω among all matrices Ω. This result is new, but it is such a natural application of the theory that we include it here rather than in a paper.

§§10-11 discuss the theory of McKean-vanMoerbeke, which describes "all" the differential identities satisfied by hyperelliptic theta functions, and especially the Matveev-Iits formula giving a solution of Kd V. We present the Adler-Gel'fand-Manin-et-al description of Kd V as a completely-integrable dynamical system in the space of pseudo-differential operators.

The second part of the chapter takes up general jacobian theta functions (i.e., $\vartheta(\vec{z},\Omega)$ for Ω the period matrix of an arbitrary Riemann surface). The fundamental special property that all such ϑ's have is expressed by the "trisecant" identity, due to John Fay (Theta functions on Riemann Surface, Springer Lecture Notes 352), and the Chapter is organized around this identity:

§1 is a preliminary discussion of the "Prime form" $E(x,y)$ — a gadget defined on a compact Riemann surface X which vanishes iff $x = y$.

§2 presents the identity.

§§3-4 specialize the identity and derive the formulae for solutions of the KP equation (in general) and KdV, Sine-Gordan (in the hyperelliptic case).

§5 is only loosely related, but I felt it was a mistake not to include a discussion of how algebraic geometry describes and explains the soliton solutions to KdV as limits of the theta-function solutions when g of the 2g cycles on X are "pinched".

The third part of the chapter by Hiroshi Umemura derives the formula mentioned above for the roots of an arbitrary algebraic equation in terms of hyperelliptic theta functions and hyperelliptic integrals.

There are two striking unsolved problems in this area: the first, already mentioned, is to find the differential identities in $\vec{z}$ satisfied by $\vartheta(\vec{z},\Omega)$ for general Ω. The second is called the "Schottky problem": to characterize the jacobians X_Ω among all abelian varieties, or to characterize the period matrices Ω of Riemann surfaces among all Ω. The problem can be understood in many ways: (a) one can seek geometric properties of X_Ω and especially of the divisor Θ of zeroes of $\vartheta(\vec{z},\Omega)$ to characterize jacobians or (b) one can seek a set of modular forms in Ω whose vanishing implies comes from a Riemann surface. One can also simplify the problem by (a) seeking only a generic characterization: conditions that define the jacobians plus possibly some other irritating components, or (b) seeking identities involving auxiliary variables: the characterization then says that X_Ω is a jacobian iff $\exists$ choices of the auxiliary variables such that the identities hold. In any case, as this book goes to press substantial progress is being made on this exciting problem. I refer the reader to forthcoming papers:

E. Arbarello, C. De Concini, On a set of equations characterizing Riemann matrices,

T. Shiota, Soliton equations and the Schottky problem,

B. van Geemen, Siegel modular forms vanishing on the moduli space of curves,

G. Welters, On flexes of the Kummer varieties.

The material for this book dates from lectures at the Tata Institute of Fundamental Research (Spring 1979), Harvard University (fall 1979) and University of Montreal (Summer 1980). Unfortunately, my purgatory as Chairman at Harvard has delayed their final preparation for 3 years. I want to thank many people for help and permissions, especially Emma Previato for taking notes that are the basis of Ch. IIIa, Mike Stillman for taking notes that are the basis of Ch. IIIb, Gert Sabidusi for giving permission to include the Montreal section here rather than in their publications, and S. Ramanathan for giving permission to include the T.I.F.R. section here. Finally, I would like to thank Birkhauser-Boston for their continuing encouragement and meticulous care.

§0. Review of background in algebraic geometry.

We shall work over the complex field $\mathbb{C}$.

Definition 0.1. An affine variety is a subset $X \subset \mathbb{C}^n$, defined as the set of zeroes of a prime ideal $\mathfrak{p} \subset \mathbb{C}[X_1,\ldots,X_n]$; $X = \{x \in \mathbb{C}^n \mid f(x) = 0 \text{ for all } f \in \mathfrak{p}\}$[1]. X will sometimes be denoted by $V(\mathfrak{p})$ or by $V(f_1,\cdots,f_k)$ if $f_1,\cdots,f_k$ generate $\mathfrak{p}$.

A morphism between two affine varieties X,Y is a polynomial map $f\colon X \to Y$, i.e., if $(X_1,\cdots,X_n) \in X$, then the point $f(X_1,\ldots,X_n)$ has coordinates $Y_i = f_i(X_1,\ldots,X_n)$, where $f_i \in \mathbb{C}[X_1,\ldots,X_n]$; following this definition, we will identify isomorphic varieties, possibly lying in different (dimensional) $\mathbb{C}^n$'s.

A variety is endowed with several structures:

a) 2 topologies; the "complex topology", induced as a subspace of $\mathbb{C}^n$, with a basis for the open sets given by $\{(x_1,\ldots,x_n) \mid |x_i - a_i| < \varepsilon, \text{ all } i\}$, and the "Zariski topology" with basis $\{(x_1,\ldots,x_n) \mid f(x) \neq 0\}$, $f \in \mathbb{C}[X_1,\ldots,X_n]$.

b) the affine ring $R(X) = \mathbb{C}[X_1,\ldots,X_n]/\mathfrak{p}$, which can be viewed as a subring of the ring of $\mathbb{C}$-valued functions on X since $\mathfrak{p}$ is the kernel of the restriction homomorphism defined on $\mathbb{C}$-valued polynomial functions on $\mathbb{C}^n$, by the Nullstellensatz.

c) the function field $\mathbb{C}(X)$, which is the field of fractions of $R(X)$; the local rings $\mathcal{O}_x$ and $\mathcal{O}_{Y,X}$, where x is a point, Y a subvariety of X, defined by $\mathcal{O}_x = \{f/g \mid f,g \in R(X) \text{ and } g(x) \neq 0\}$, with maximal ideal $m_x = \{f/g \in \mathcal{O}_x \mid f(x) = 0\}$, $\mathcal{O}_{Y,X} = \{f/g \mid f,g \in R(X), g \not\equiv 0 \text{ on } Y\} = R(X)_{\mathfrak{q}}$[2] if $Y = Y(\mathfrak{q})$;

1) If a polynomial $f \in \mathbb{C}[X_1,\cdots,X_n]$ is zero at every point of V then $f \in \mathfrak{p}$; this is Hilbert's Nullstellensatz.
2) We denote by $A_{\mathfrak{q}}$ the localization of a domain A with respect to its prime ideal $\mathfrak{q}$, $A_{\mathfrak{q}} = \{a/b \mid a,b \in A,\ b \notin \mathfrak{q}\}$.

(notice: if $x \in Y$, $R(X) \subset \mathcal{O}_x \subset \mathcal{O}_{Y,X} \subset \mathbb{C}(X)$); the structure sheaf $\mathcal{O}_X$, subsheaf of the constant sheaf $U \longmapsto \mathbb{C}(X)$, which assigns to any Zariski-open subset U of X the ring $\bigcap_{x \in U} \mathcal{O}_x = \Gamma(U, \mathcal{O}_X)$; and a dimension given by $\dim X = \text{tr.d}_{\mathbb{C}}\mathbb{C}(X)$. dim X is related to the Krull dimension of $\mathcal{O}_{Y,X}$ (maximum length of a chain of prime ideals), by:

Proposition 0.2. $\dim X - \dim Y = \text{Krull dim. } \mathcal{O}_{Y,X}$.

d) the Zariski tangent-space at $x \in X$, which can be defined in a number of equivalent ways:

$T_{X,x}$ = vector space of derivations $d: R(X) \longrightarrow \mathbb{C}$ centered at x (i.e., satisfying the product rule $d(fg) = f(x)dg+g(x)df$); or

$T_{X,x} = (\mathfrak{m}_x/\mathfrak{m}_x^2)^\vee$, the space of linear functions on $\mathfrak{m}/\mathfrak{m}^2$; or

$T_{X,x}$ = the space of n-tuples $(\dot{x}_1,\cdots,\dot{x}_n)$ such that for all $f \in \mathfrak{p}$, $f(x_1+\varepsilon\dot{x}_1,\cdots,x_n+\varepsilon\dot{x}_n) \equiv 0 \bmod \varepsilon^2$,

where from a derivation d a linear function $\ell(X_i-x_i) = dX_i$ and an n-tuple $(\dot{x}_1,\cdots,\dot{x}_n)$ with $dX_i = \dot{x}_i$ are obtained; this sets up the bijection. This vector is also written customarily as $\sum_{i=1}^{n} (\dot{x}_i)\partial/\partial x_i$;

Proposition 0.3. $\exists$ a non-empty Zariski open subset $U \subset X$ such that $\text{tr.d.}_{\mathbb{C}}\,\mathbb{C}(X) = \dim T_{X,x}$ for all $x \in U$; if $x \notin U$, then $\dim T_{X,x} > \dim X$.

U is called the set of "smooth" points of X, X-U the "singular locus". It can be shown from this proposition that U (with the complex topology) is locally homeomorphic to $\mathbb{C}^d$, where d is $\text{tr.deg.}_{\mathbb{C}}\mathbb{C}(X)$.

Lemma 0.4. For any $x \in X$, $\exists$ a fundamental system of Zariski neighborhoods U of x such that U is isomorphic to an affine variety.

In fact, for any $f \in R(X)$ such that $f(x) \neq 0$, $U_f = \{y \in X \mid f(y) \neq 0\}$ is a neighborhood of x and if $R_X = \mathbb{C}[X_1,\ldots,X_n]/\mathfrak{P}$, then U_f is isomorphic to the subvariety of $\mathbb{C}^{n+1}$ is defined by the ideal $(\mathfrak{P}, X_{n+1}f(X_1,\ldots,X_n)-1$; the isomorphism is realized by

$$(x_1,\cdots,x_n) \longmapsto (x_1,\cdots,x_n, f(x_1,\cdots,x_n)^{-1}).$$

But we need a more subtle definition of morphism from an open set to an affine variety.

Definition 0.5. $f\colon \underset{X}{U \underset{\text{open}}{\cap}} \longrightarrow Y$ is a morphism if (equivalently):

(1) for any $g \in R(Y)$, thought of as a complex-valued function on Y, $g \circ f \in \Gamma(U, \mathcal{O}_X)$

(2) $\exists$ $g_{ik}, h_k \in \mathbb{C}[X_1,\cdots,X_m]$ such that for any $(x_1,\cdots,x_n) \in U$ there is a suitable k such that $h_k(x) \neq 0$, and the i-th coordinate of $f(x_1,\cdots,x_n)$ is given by $\dfrac{g_{ik}(x_1,\cdots,x_n)}{h_k(x_1,\cdots,x_n)}$ whenever $h_k(x) \neq 0$.

(n.b. there may not exist a single expression $f(X_1,\cdots,X_m)_i = \dfrac{g_i(X_1,\cdots,X_n)}{h(X_1,\cdots,X_n)}$, with $h^{-1} \in \Gamma(U, \mathcal{O}_X)$.

Theorem 0.6 (Weak Zariski's Main Theorem). If $f\colon X \longrightarrow Y$ is an injective morphism between affine varieties of the same dimension and Y is smooth, then f is an isomorphism of X with an open subset of Y.

The product of affine varieties is categorical, i.e., given $X \subset \mathbb{C}^n$ and $Y \subset \mathbb{C}^m$ affine varieties, i) $X \times Y$ is an affine variety (in $\mathbb{C}^{n+m}$), ii) the projections are morphisms, iii) if Z is an affine variety and morphisms $Z \longrightarrow X$, $Z \longrightarrow Y$ are given, then there is a unique morphism $Z \longrightarrow X \times Y$ making a commutative diagram

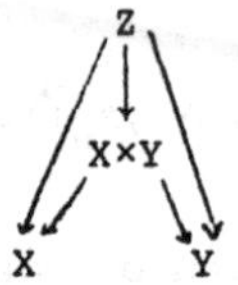

Definition 0.7. A variety in general is obtained by an atlas of affine varieties: $X = \bigcup_{\alpha \in S} X_\alpha$, S a finite set, $X_\alpha \subset \mathbb{C}^{n_\alpha}$, glued by isomorphisms

$$\begin{array}{ccc} U_{\alpha,\beta} & \subset & X_\alpha \\ \phi_{\alpha\beta} \downarrow \uparrow \phi_{\alpha\beta}^{-1} & & \\ U_{\beta,\alpha} & \subset & X_\beta \end{array}$$

(where $U_{\alpha,\beta}$ is a nonempty Zariski-open subset of X_α), such that one of the equivalent (separation) conditions holds:

(1) X is Hausdorff in the "complex topology" (a subset of X being open in the complex topology if and only if its intersections with X_α are open for all α's)

(2) the graph of $\phi_{\alpha\beta}, \Gamma_{\alpha\beta} \subset X_\alpha \times X_\beta$ is Zariski-closed.

(3) for any valuation ring $R \subset \mathbb{C}(X) = \mathbb{C}(X_\alpha)$ (any α, for the function field of $U_f \subset U_{\alpha\beta}$ coincides with that of X_α, hence $\phi_{\alpha\beta}$ identifies $\mathbb{C}(X_\alpha)$ and $\mathbb{C}(X_\beta)$) there is at most one point $x \in X$

such that $R > \mathcal{O}_x$ (R "dominates" $\mathcal{O}_x$, or R is "centered" at x, i.e., $R \supset \mathcal{O}_x$ and $m_R \supset m_x$).

(4) for all affine varieties Y and morphisms $f,g: Y \longrightarrow X$, the set $\{y \in Y \mid f(y) = g(y)\}$ is (Zariski) closed in Y.

Such an X carries:

(a') 2 topologies (the complex and the Zariski; as with the complex topology, a subset of X is Zariski-open if and only if its intersection with all the X_α's is Zariski-open)

(c') the function field $\mathbb{C}(X)$; the local rings[3)] $\mathcal{O}_x = \{f/g \mid f,g \in R(X_\alpha),\ g(x) \neq 0\}$ if $x \in X_\alpha$; the structure sheaf $U \longmapsto \bigcap_{x \in U} \mathcal{O}_x$

(d') the Zariski tangent space $T_{x,X} = T_{x,X_\alpha}$ if $x \in X_\alpha$.

$f: X \longrightarrow Y$ is a morphism between two varieties if the restriction $\text{res}\ f: U_\alpha \cap f^{-1}(V_\beta) \longrightarrow V_\beta$ is a morphism for all α,β's or, equivalently, if for any open set $U \subset Y$ and $g \in \Gamma(U,\mathcal{O}_Y)$, $g \circ f \in \Gamma(f^{-1}U,\mathcal{O}_X)$ is satisfied.

<u>Key example</u>. Projective varieties, defined by homogeneous ideals $\mathfrak{p} \subset \mathbb{C}[X_0,\ldots,X_n]$, as

$$V(\mathfrak{p}) = \{(x_0,\cdots,x_n) \in \mathbb{P}^n \mid f(x) = 0 \text{ for all } f \in \mathfrak{p}\};$$

an atlas is given by $V(\mathfrak{p})_i = \{p \in V(\mathfrak{p}) \mid X_i(p) \neq 0\}$.

---

3) A variety X can even be defined as a set of local rings $\{\mathcal{O}_x\}$ with the same fraction field $\mathbb{C}(X)$. Then the topology on X is defined as follows - for each $f \in \mathbb{C}(X)$, let U_f be the set of the local rings containing f.

The product of varieties is again a variety; we take $(U\times V)_{\Sigma f_i g_i}$ to be a basis for the open sets in $X\times Y$, where U,V are open subsets of X,Y isomorphic to affine varieties, $f_i \in \Gamma(U,\mathcal{O}_X)$, $g_i \in \Gamma(V,\mathcal{O}_Y)$ and $(U\times V)_{\Sigma f_i g_i}$ is the set of points $(x,y) \in U\times V$ such that $\Sigma f_i(x)g_i(y) \neq 0$.

The product of projective varieties is again a projective variety, for instance the map $(x_i,y_j) \longmapsto (x_i y_j)$ embeds $\mathbb{P}^n \times \mathbb{P}^m$ into $\mathbb{P}^{(n+1)(m+1)-1}$ and the image is given on the affine pieces $(\mathbb{P}^{n+m+nm})_{X_{hk}}$ by the equations $s_{ij} = s_{ih}s_{kj}$ for all $i \neq h$ and $j \neq k$, where $s_{ij} = X_{ij}/X_{hk}$.

Definition 0.8. A variety X is complete (or proper) if one of the following equivalent condition holds:

(1) X is compact in the complex topology

(2) $\exists$ a surjective birational morphism $f: X' \longrightarrow X$, X' projective

(3) for all valuation rings $R \subset \mathbb{C}(X)$, $\exists\, x \in X$ such that $R > \mathcal{O}_x$

(4) for all varieties Y, $Z \subset X\times Y$ closed, $pr_2 Z$ is closed in Y.

A subvariety of a variety X is an irreducible locally closed subset Y of X; the variety structure is given by the sheaf $\mathcal{O}_Y$ which assigns to any open subset V of Y the ring

$$\Gamma(V,\mathcal{O}_Y) = \left\{ \mathbb{C}\text{-valued functions } f \text{ on } V \;\middle|\; \begin{array}{l} \forall x \in V,\ \exists \text{ a neighborhood } U \\ \text{of } x \text{ in } X \text{ and a function} \\ f_1 \in \Gamma(U,\mathcal{O}_X) \text{ such that} \\ f = \text{restriction to } U\cap V \text{ of } f_1 \end{array} \right\}$$

So, any open subset of X is a subvariety; but a subvariety which is a complete variety must be closed.

Divisors and linear systems.

The theory of divisors is based on a fundamental result of Krull.

0.9. If R is a noetherian integrally closed integral domain, then

a) for all $\mathfrak{p} \subset R$, $\mathfrak{p}$ minimal prime ideal, $R_{\mathfrak{p}}$ is a discrete valuation ring.

b) $R = \bigcap_{\mathfrak{p} \text{ min. prime}} R_{\mathfrak{p}}$.

Thus if $\mathrm{ord}_{\mathfrak{p}}$ = valuation attached to $R_{\mathfrak{p}}$, and K is the fraction field of R, we get an exact sequence:

$$1 \longrightarrow R^* \longrightarrow K^* \longrightarrow \left[\begin{array}{l}\text{free abel. group}\\ \text{on min. prime ideals}\end{array}\right]$$

$$f \longmapsto \sum_{\mathfrak{p}} \mathrm{ord}_{\mathfrak{p}} f \cdot [\mathfrak{p}] = (f) \quad .$$

Let $\mathfrak{p}_1,\dots,\mathfrak{p}_n$ be the primes occurring positively in (f),

$\mathfrak{p}_2',\dots,\mathfrak{p}_m'$ " " " negatively in (f), then

Corollary 0.10. For all prime ideals $\mathfrak{p}$ in R,

$$f \in R_{\mathfrak{p}} \iff \mathfrak{p} \not\supset \mathfrak{p}_i' \quad \text{any } i$$

$$f^{-1} \in R_{\mathfrak{p}} \iff \mathfrak{p} \not\supset \mathfrak{p}_i \quad \text{any } i, \text{ hence}$$

$$\left(\begin{array}{l}\text{neither } f \text{ or } f^{-1} \text{ are in } R_{\mathfrak{p}}\\ \text{"}f \text{ is indeterminate at } \mathfrak{p}\text{"}\end{array}\right) \iff \mathfrak{p} \supset \mathfrak{p}_i + \mathfrak{p}_j' \quad \text{for some } i,j$$

(in particular, if f is indeterminate at $\mathfrak{p}$, then $\mathfrak{p}$ is not a minimal prime ideal).

We will apply Krull's result to the following geometrical situation:

Theorem 0.11: If $X = \bigcup X_\alpha$ is a smooth variety, then R_{X_α} is integrally closed, the minimal primes $\mathfrak{p}$ in R_{X_α} are the codimension one (closed) subvarieties Y of X which meet X_α, and $(R_{X_\alpha})_{\mathfrak{p}} = \mathcal{O}_{Y,X}$.

Idea of the proof: for all points $P \in X_\alpha$, the hypothesis of being smooth means $\dim \mathfrak{m}_P/\mathfrak{m}_P^2 = \dim X = \text{Krull-dim. } \mathcal{O}_P$, i.e., $\mathcal{O}_P$ is "regular" (this can be taken as a definition). One proves that a regular local ring is integrally closed, hence $\mathcal{O}_P$ is integrally closed. Since, for any affine variety, $R_{X_\alpha} = \bigcap_{P \in X_\alpha} \mathcal{O}_P$ [5], R_{X_α} is integrally closed. The rest of the statement follows from the:

Lemma 0.12. A (closed) subvariety Y of Z is maximal $\Longleftrightarrow$ dim Y = dim Z-1.

(This follows from (o.2), or else can be used to prove (o.2).)

Thus the map $f \longmapsto (f)$ defines a homomorphism

$$\mathbb{C}(X)^* \longrightarrow \text{Div } X = \begin{bmatrix} \text{free abel. group} \\ \text{on codim. 1 subvar.} \end{bmatrix}$$

Elements of Div X are called divisors on X and 2 divisors D_1, D_2 are called linearly equivalent (written $D_1 \equiv D_2$) if $D_1 - D_2 = (f)$, some $f \in \mathbb{C}(X)^*$.

The corollary 0.10 has the following geometrical meaning: for any $f \in \mathbb{C}(X)^*$, set $(f) = (f)_0 - (f)_\infty$ with $(f)_0$ (zero-divisor) and $(f)_\infty$ (pole-divisor) both positive divisors, and let, for any divisor $D = \sum n_i Y_i$, $\text{supp } D = \cup Y_i$; then

5) If $x/y \in \bigcap_{P \in X_\alpha} \mathcal{O}_P$, consider the ideal $A = \{z \in R_{X_\alpha} \mid z \cdot \frac{x}{y} \in R_{X_\alpha}\}$; since $x/y \in \mathcal{O}_P$, x/y can be written w/z, with $w \in R_{X_\alpha}$, $z \in R_{X_\alpha} - M_P$, so $A \not\subset M_P$. Therefore A is not contained in any maximal ideal, so $A = R_{X_\alpha}$. This means that $1 \in A$, i.e., $\frac{x}{y} \in R_{X_\alpha}$.

$$f \in \mathcal{O}_P \iff P \notin \operatorname{supp}(f)_\infty$$

$$f^{-1} \in \mathcal{O}_P \iff P \notin \operatorname{supp}(f)_0$$

$$f \text{ is indeterminate at } P \iff P \in \operatorname{supp}(f)_0 \cap \operatorname{Supp}(f)_\infty.$$

Moreover, if X is a smooth affine variety of dimension 1 with affine ring R, then R is a Dedekind domain, so all its ideals are products of prime ideals. If $f \in R$, let:

$(f) = \Sigma n_i Y_i$ where Y_i corresponds to the prime ideal $\mathfrak{p}_i$ in R.

Then:

Corollary 0.14.

$$f \cdot R = \prod_i \mathfrak{p}_i^{n_i}.$$

We define $\mathrm{Div}^+(X)$ to be the semi-group in Div(X) of divisors with only positive coefficients.

We define Pic(X) as the cokernel:

$$\mathbb{C}(X)^* \longrightarrow \mathrm{Div}\, X \xrightarrow{\pi} \mathrm{Pic}(X) \longrightarrow 0 ,$$

i.e., as the obstruction to finding rational functions with given zeroes and poles. Elements of Pic(X) are called divisor classes.

Example. $\mathrm{Pic}(\mathbb{P}^n) = \mathbb{Z}$. In fact, any hypersurface is given by the zeroes of a homogeneous polynomial. The degree of a divisor $D = \Sigma n_i Y_i$ is defined by $\deg D = \Sigma n_i \deg Y_i$ where $\deg Y_i$ is the degree of the irreducible homogeneous polynomial defining it. Then any divisor of degree zero comes from a rational function, and degree gives an isomorphism $\mathrm{Pic}(\mathbb{P}^n) \xrightarrow{\sim} \mathbb{Z}$.

Suppose D is a positive divisor; we define the vector space

$$\mathcal{L}(D) = \{f \in \mathbb{C}(X)^* \mid (f)+D \geq 0\} \cup \{0\}.$$

Note: The condition $(f)+D \geq 0$ is equivalent to $(f)_\infty \leq D$ (the poles of f are bounded by D). Note that $\mathcal{L}(D)$ is a sub-vector space of $\mathbb{C}(X)$.

Lemma 0.15. If X is proper, $\dim \mathcal{L}(D) < \infty$ and for all $f \in \mathbb{C}(X)^*$ $(f) = 0$ if and only if $f \in \mathbb{C}^*$.

In this case, we form the associated projective space $\mathbb{P}(\mathcal{L}(D))$ of one-dimensional subspaces of $\mathcal{L}(D)$ and note:

$$\begin{array}{ccccc} \mathbb{P}(\mathcal{L}(D)) & \cong & \left[\pi^{-1}(\pi D) \cap \mathrm{Div}^+(X)\right] & = & \left[\begin{array}{l}\text{fibre through D of} \\ \mathrm{Div}^+(X) \xrightarrow{\pi} \mathrm{Pic}(X)\end{array}\right] \\ \Psi & & \Psi & & \\ \text{line}\{\alpha.f \mid \alpha \in k\} & \longleftrightarrow & \text{divisor } (f)+D & & \end{array}$$

These projective spaces and their linear subspaces are the so-called "linear systems" of divisors. $\mathbb{P}(\mathcal{L}(D))$ is denoted $|D|$.

If $L \subset |D|$ is a linear subspace of dimension k, set

$$B(L) = \bigcap_{E \in L} \mathrm{Supp}\, E, \text{ the "base locus" of } L.$$

The fundamental construction associated to linear systems is the map

$$\varphi_L : (X - B(L)) \longrightarrow L^\vee,$$

where $L^\vee$ is the projective space of hyperplanes in L, given by

$$x \longmapsto [\text{hyperplane in L consisting of the } E \in L \text{ s.t. } x \in \mathrm{Supp}\, E]$$

φ_L is a morphism. To prove this and to describe φ_L explicitly, let's choose a projective basis of L, i.e., k+1 points which are not contained in a hyperplane:

$$E,\ E+(f_1), E+(f_2), \cdots, E+(f_k).$$

Set $f_o = 1$; the map

$$x \longmapsto (f_o(x), \cdots, f_k(x))$$

is defined on the open set X-Supp E since the poles of f_i are all contained in Supp E; it coincides with φ_L on X-Supp E, as we see if we let coordinates on L be $c_o, \cdots, c_k$ and note: for $x \notin$ Supp E,

$$x \in \operatorname{Supp}\Big(E + \big(\sum_{i=0}^{k} c_i f_i\big)\Big) \iff \sum_{i=0}^{k} c_i\ f_i(x) = 0.$$

hence $\varphi_L(x)$ = hyperplane in L with coefficients $f_0(x), --, f_k(x)$
= pt. of $L^\vee$ with homogeneous coordinates $f_0(x), ---, f_k(x)$.

§1. Divisors on hyperelliptic curves.

Given a finite number of distinct elements $a_i \in \mathbb{C}$, $i \in S$, let $f(t) = \prod_{i\in S}(t-a_i)$. We form the plane curve C_1 defined by the equation

$$s^2 = f(t).$$

The polynomial $s^2-f(t)$ is irreducible, so $(s^2-f(t))$ is a prime ideal, and C_1 is a 1-dimensional affine variety in $\mathbb{C}^2$. In fact, C_1 is smooth. To prove this, we will calculate the dimension of the Zariski-tangent space at each point, i.e., the space of solutions $(\dot{s},\dot{t}) \in \mathbb{C}^2$ to the equation

$$(s+\varepsilon\dot{s})^2 = \prod(t+\varepsilon\dot{t}-a_i) \bmod \varepsilon^2 \qquad \text{for } (s,t) \in C_1.$$

That is equivalent to the equation

$$2s\dot{s} = \dot{t}\,.\sum_{j\in S}\prod_{i\neq j}(t-a_i);$$

if $s \neq 0$, the solutions are all linearly dependent since $\dot{s} = \frac{\dot{t}}{2s}\left(\sum_{j\in S}\prod_{i\neq j}(t-a_i)\right)$; if $s = 0$, we get from the equation of the curve $\prod_{i\in S}(t-a_i) = 0$, hence $t = a_i$ for some i; thus $0 = \dot{t}\,.\prod_{j\neq i}(a_i-a_j)$, so $\dot{t} = 0$. Thus at all points, the Zariski tangent space is one-dimensional.

We add points at infinity by introducing a second chart:

$$C_2: \qquad s'^2 = \prod_{i\in S}(1-a_it') \qquad \text{if } \#S = 2k$$

$$s'^2 = t'.\prod_{i\in S}(1-a_it') \qquad \text{if } \#S = 2k-1,$$

glued by the isomorphism $t' = \frac{1}{t}$

$$s' = \frac{s}{t^k}$$

between the open sets $t \neq 0$ of C_1 and $t' \neq 0$ of C_2.
The points at ∞ of C_1 are:

∞_1, ∞_2 given by $t' = 0$, $s' = \pm 1$ if #S even

∞ " " $t' = 0 = s'$ if #S odd.

On the resulting variety C we can define a morphism π: $C \longrightarrow \mathbb{P}^1$. Let t and $t' = 1/t$ be affine coordinates in $\mathbb{P}^1$; then define π by

$$(s,t) \longmapsto t, \text{ on the chart } C_1,$$
$$(s',t') \longmapsto t', \text{ on the chart } C_2.$$

π is 2:1 except over the set B of the "branch points" consisting in the a_i's, and ∞ in the case #S odd. The number of branch points is therefore an even number 2k in both cases. Topologically C is a surface with k-1 handles, so we say that it is of genus $g = k-1$; this is called the <u>genus</u> of the curve. This is usually visualized by defining 2 continuous functions $+\sqrt{f(t)}$, $-\sqrt{f(t)}$ for $t \in \mathbb{P}^1-$(k"cuts") and reconstructing C by glueing the 2 open pieces of C defined by $s = +\sqrt{f(t)}$ and $s = -\sqrt{f(t)}$:

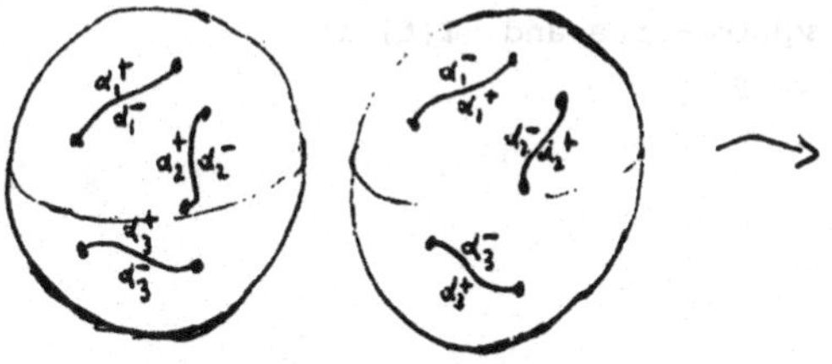

k disjoint cuts on each copy of $\mathbb{P}^1$

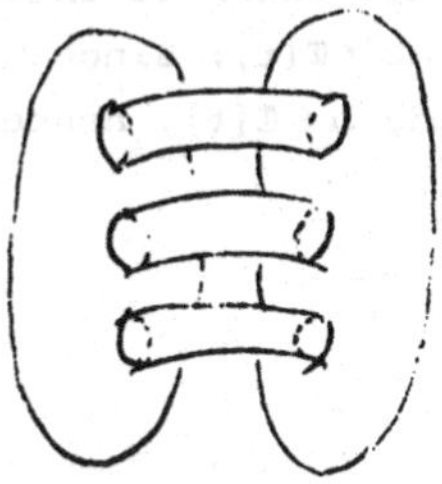

glueing α_i^+'s, α_i^-'s

Since C is smooth, the affine rings of C_1 and C_2 are Dedekind domains[1)], and their local rings $\mathcal{O}_x$ are discrete valuation rings.

$$\iota: \quad (s,t) \longmapsto (-s,t)$$

is an automorphism of C, that flips the sheets of the covering, hence is an involution, with the set of orbits $C/\{\pm 1\} \cong \mathbb{P}^1$. $\pi^{-1}(B)$ is the set of fixed points of ι.

We want to prove that C is actually a projective variety.

Let

$$D = \begin{cases} k(\infty_1 + \infty_2) & \text{if } \#S \text{ is even} \\ 2k\infty & \text{if } \#S \text{ is odd.} \end{cases}$$

Lemma 1.1. $1,t,t^2,\cdots,t^k,s$ is a basis for the vector space $\mathcal{L}(D)$.

---

1) We already know that the tangent space to the curve at each point has the right dimension, in each of the two affine pieces; but it's also easy to see directly that $\mathbb{C}[t,s]/(s^2-\Pi(t-a_i))=R$ is integrally closed, the reason being that $\Pi(t-a_i)$ is a square-free discriminant over the U.F.D. $\mathbb{C}[t]$. If we let σ be the automorphism which sends (s,t) to $(-s,t)$, then the general element of the quotient field of R is $a+bs$, with $a,b\in\mathbb{C}(t)$, and for all $a+bs$ integral over R, $(a+bs)+\sigma(a+bs)=2a$ and $(a+bs)\cdot\sigma(a+bs) = a^2-b^2d$ are in $\mathbb{C}(t)$ and are integral over $\mathbb{C}[t]$, which is integrally closed. Thus $2a\in\mathbb{C}[t]$, $a^2-db^2\in\mathbb{C}[t]$, so $db^2\in\mathbb{C}[t]$; since d is square-free and $\mathbb{C}[t]$ is U.F.D. we conclude $b\in\mathbb{C}[t]$, hence $a+bs\in R$.

<u>Proof</u>: The function field of X, $\mathbb{C}(t)[\sqrt{\Pi(t-a_i)}]$, has an involution over $\mathbb{C}(t)$, that interchanges ∞_1, ∞_2, or fixes the point ∞, hence sends $\mathcal{L}(D)$ into itself. Thus $\mathcal{L}(D)$ splits into the sum of the +1 and -1 eigenspaces of ι,

$$\mathcal{L}(D) = [\mathcal{L}(D) \cap \mathbb{C}(t)] \oplus [\mathcal{L}(D) \cap s\mathbb{C}(t)].$$

If $h(t) \in \mathcal{L}(D) \cap \mathbb{C}(t)$, since it has no poles for finite values of t, then it must be a polynomial in t, $h \in \mathbb{C}[t]$. On the other hand, in the case #S even, the maximum ideal of $\mathcal{O}_{\infty_1} = R(C_2)_{(t',s'-1)}$ is generated by t' since the equation of the curve gives $s'-1 = ((s'+1)^{-1} \cdot (\Pi(1-t'a_i)-1) \in (t')R(C_2)_{(t',s'-1)}$; in the case #S odd the max. ideal of $\mathcal{O}_\infty = R(C_2)_{(t',s')}$ is generated by s', since $t' = s'^2 (\Pi(1-t'a_i))^{-1}$; thus

$v_{\infty_1}(t') = -v_{\infty_1}(t) = 1$ (similarly $v_{\infty_2}(t') = 1$), or $v_\infty(t') = -v_\infty(t) = 2$,

$$\text{i.e., } (t)_\infty = \begin{cases} \infty_1 + \infty_2, & \#S \text{ even} \\ 2\infty, & \#S \text{ odd} \end{cases}$$

So in order for $(h)_\infty$ to be $\leq D$ we must have $\deg h \leq k$.

Now consider $h(t) \in \mathcal{L}(D) \cap s\mathbb{C}(t)$, $h = sg(t)$ with $g(t) \in \mathbb{C}(t)$; g may only have poles in C_1 where s has zeroes, i.e., in the set $\{P_i = (0,a_i)\}$. The order of vanishing of s at P_i is 1 and that of $(t-a_i)$ is 2, since the max. ideal in $\mathcal{O}_{P_i}$ is generated by s and $(t-a_i)$ and $(t-a_i) = s^2 \cdot (\prod_{j\neq i} (t-a_j))^{-1}$. That prevents $g(t) \in \mathbb{C}(t)$

from having a pole at P_i, because the product $sg(t)$ would still have a pole at P_i. Thus the only poles of $g(t)$ must be at ∞, i.e., $g(t)$ must be a polynomial in t; but now

$$(s)_\infty = (s't^k)_\infty = \begin{cases} k(\infty_1+\infty_2) & \#S \text{ even} \\ (2k-1)\infty & \#S \text{ odd} \end{cases}$$

hence $g(t)$ must be constant in order to have $D+(sg(t)) \geq 0$. This proves the lemma.

Now a projective base for $|D|$ is $D, D+(t), \cdots, D+(t^k)$, $D+(s)$; since $(t)_\infty = \begin{cases} \infty_1+\infty_2 \\ 2\infty \end{cases}$ we have $D+(t^k) = k\cdot(t)_0$ = either $k(O_1+O_2)$ or $2kO$ where O is a branch point in the 2nd case, and where O_1, O_2 are the two points in the fiber over the point $t = 0$ of $\mathbb{P}^1$ in the 1st case. Hence $|D|$ contains D and $k\cdot(t)_0$ whose supports are disjoint, hence $B|D|$ is empty.

Thus explicitly, corresponding to $|D|$, we get $\varphi_{|D|}: C \longrightarrow \mathbb{P}^{k+1}$

by $(s,t) \longmapsto (1,t,t^2,\cdots,t^k,s)$ on $X - \text{Supp } D = C_1$,

and $(s',t') \longmapsto (t'^k,\ldots\ldots,t',1,s')$ on C_2.

Note that these 2 maps do agree on the overlap: $(1,t,t^2,\cdots t^k,s) \sim \frac{1}{t^k}(1,t,\ldots t^k,s) = (t'^k,\cdots,t',1,s')$.

This map, which is an isomorphism of C with its image, makes C into a projective curve.

Remark. If $\psi: \mathbb{P}^1 \longrightarrow \mathbb{P}^1$ is a linear fractional transformation $\left(\text{i.e., } \psi(t) = \frac{at+b}{ct+d}, \begin{pmatrix} a & b \\ c & d \end{pmatrix} \in SL(2,\mathbb{C})\right)$, then it is not hard to check that the two hyperelliptic curves whose sets of branch points are, respectively, B and $\psi(B)$ are isomorphic. So we can henceforth assume that #S is always odd by sending one branch point to ∞.

Our aim is to describe a variety of divisors on C, and from this the Jacobian variety of C; the idea of this construction is due originally to Jacobi and appeared in

"Uber eine neue Methode zur Integration der hyperelliptischen Differentialgleichungen und über die rationale Form ihrer vollständigen algebraischen Integralgleichungen" Crelle, 32, 1846.

Let's consider the subset $\mathrm{Div}^{\nu}(C)$ of Div(C) given by all the divisors of degree ν : $\mathrm{Div}(C) = \coprod_{\nu \in \mathbb{Z}} \mathrm{Div}^{\nu}(C)$, and inside the set of positive ones $\mathrm{Div}^{+,\nu}$, those with the following property:

$$\mathrm{Div}^{+,\nu}(C) \supset \mathrm{Div}_o^{+,\nu}(C) = \left\{ D \in \mathrm{Div}^{+,\nu}(C) \;\middle|\; \begin{array}{l} \text{if } D = \sum_{i=1}^{\nu} P_i, \text{ then } P_i \neq \infty \text{ all } i \\ \text{and } P_i \neq \iota(P_j) \text{ all } i \neq j \end{array} \right\}$$

Our basic idea is to associate to $D \in \mathrm{Div}_o^{+,\nu}(C)$ three polynomials:

(a) $U(t) = \prod_{i=1}^{\nu} (t - t(P_i))$, monic of degree ν $\big(t(P_i)$ is the value of t at $P_i\big)$

(b) If the P_i's are distinct, let

$$V(t) = \sum_{i=1}^{\nu} s(P_i) \frac{\prod_{j\neq i} (t-t(P_j))}{\prod_{j\neq i} (t(P_i)-t(P_j))} .$$

V(t) is the unique polynomial of degree $\leq \nu-1$ such that

$$V(t(P_i)) = s(P_i), \qquad 1 \leq i \leq \nu .$$

If P_i has positive multiplicity in D, then we want to "approximate the function $\sqrt{f(t)}$ up to the order $m_D(P_i)$", and in order to do that we let

$$V(t) = \left[\begin{array}{l} \text{the unique polynomial of degree} \leq \nu-1 \text{ such that, if} \\ m_D(P_i) = n_i, \\ \left(\frac{d}{dt}\right)^j \left[V(t) - \sqrt{\prod_{\ell \in S} (t-a_\ell)}\right]\Bigg|_{t=t(P_i)} = 0 \quad \text{for } 0 \leq j \leq n_i-1 \end{array}\right].$$

By construction $f(t)-V(t)^2$ is divisible by U(t); hence

(c) Define W(t) by: $f(t)-V(t)^2 = U(t).W(t)$.

Let's assume $\nu \leq g+1$; then $\deg V(t)^2 < \deg f(t)$ and since U(t) is monic in t of degree ν, W is monic of degree $2g+1-\nu$.

Conversely, given any U,V,W such that $f-V^2 = UW$, U and W monic, having degrees as above, we get the divisor $(U)_0$ of ν points on the t-line; over each zero of U, the corresponding value of V gives a square root of f(t), i.e., a value of s (either one of $\pm\sqrt{f(t)}$); thus the divisor of the points on the curve so obtained is in $\mathrm{Div}_0^{+,\nu}$.

Remark. Given U,V,W satisfying the above equation, then in $\mathbb{C}[s,t]$:

$$s^2-f(t) = (s-V(t))\cdot(s+V(t))-U(t)\cdot W(t), \text{ hence}$$
$$(s^2-f(t)) \subset (U(t),s-V(t)),$$

and the bigger ideal defines a zero-dimensional subset of C, which is in fact supp D, or a zero-dimensional subscheme, which is D. We have now proven:

Proposition 1.2. There is a bijection between

$$\mathrm{Div}_o^{+,\nu} \quad \text{and} \quad \left\{ \begin{array}{c|l} \text{triples of polynomials} & f-V^2=UW,\ U,\ W \text{ are monic,} \\ U,V,W & \deg V \le \nu-1,\ \deg U = \nu,\ \deg W = 2g+1-\nu \end{array} \right\}$$

Notice how the bijection gives us a way to introduce coordinates into $\mathrm{Div}_o^{+,\nu}(X)$: let

$$\begin{aligned} U(t) &= t^\nu + U_1t^{\nu-1} +\dots + U_\nu \\ V(t) &= \phantom{t^\nu +{}} V_1t^{\nu-1} +\dots + V_\nu \\ W(t) &= t^{2g+1-\nu} + W_0t^{2g-\nu} +\dots + W_{2g-\nu} \end{aligned}$$

be 3 polynomials with indeterminant coefficients, and expand:

$$f - V^2 - UW = \sum_{\alpha=0}^{2g} a_\alpha(U_i,V_j,W_\ell)t^\alpha.$$

Then, taking U_i,V_j,W_k as coordinates:

$$[\text{the set of triples } (U,V,W) \text{ as above}] \cong V(a_0,\cdots,a_{2g}) \subset \mathbb{C}^{2g+1+\nu}.$$

Or else, since U and V determine W whenever the division is possible, we can write using the Euclidean algorithm:

$$f(t)-V(t)^2 = U(t).[t^{2g+1-\nu}+B_0(U_i,V_j)t^{2g-\nu}+..]+\underbrace{R_1(U_i,V_j)t^{\nu-1}+\cdots+R_\nu(U_i,V_j)}_{\text{remainder}}$$

Using only U_i,V_j as coordinates, we find:

$$[\text{the set of triples } (U,V,W) \text{ as above}] \cong V(R_1,\cdots,R_\nu) \subset \mathbb{C}^{2\nu}.$$

The structure of affine variety is the same in both cases because the morphisms:

$$(U_i,V_j,W_k) \xmapsto{\text{projection}} (U_i,V_j) \quad \text{and}$$

$$(U_i,V_j,B_k(U_i,V_j)) \longleftarrow (U_i,V_j)$$

are inverse of one another.

On the other hand, we can parametrize $\mathrm{Div}_0^{+,\nu}(C)$ by points of C, in the following way: we have a surjective map

$$C^\nu \longrightarrow\!\!\!\!\to \mathrm{Div}^{+,\nu}(C)$$

$$(P_1,\cdots,P_\nu) \longmapsto \sum P_i \;;$$

now let $(C^\nu)_0 \subset C^\nu$ be the Zariski open set defined as follows:

$$C^\nu - (C^\nu)_o = \left[\bigcup_{i=1}^{\nu} p_i^{-1}(\infty)\right] \cup \left[\bigcup_{0\le i\le j\le \ell} p_{ij}^{-1}(\Gamma)\right]$$

where $p_i\colon C^\nu \longrightarrow C$ is the i-th projection, $p_{ij}\colon C^\nu \longrightarrow C^2$ the (i,j)-th projection, $\Gamma = [\text{locus of points } (P,\iota P)]$, , the Zariski closed subset of C^2 given by the equations $s_1 = -s_2$, $t_1 = t_2$ if (s_1,t_1,s_2,t_2) are coordinates. Then everything is tied together in:

<u>Proposition 1.3.</u> <u>The equations</u> $a_0,\cdots,a_{2g}$ <u>generate a prime ideal in</u> $\mathbb{C}[U_i,V_j,W_\ell]$, <u>the variety</u> $V(a_0,\cdots,a_{2g})$ <u>is smooth and the composite map</u> $(C^\nu)_o \longrightarrow\!\!\!\!\to \mathrm{Div}_o^{+,\nu}(C) \cong V(a_0,\cdots,a_{2g})$ <u>is a surjective morphism making</u>

$$V(a_0,\cdots,a_{2g}) \cong \left[\frac{\text{orbit space for the group of permutations}}{S_\nu \text{ acting on } (C^\nu)_o}\right]$$

The proof of proposition 1.3 will consist of 2 steps.

1. In order to prove that $V(a_\alpha)$ is smooth, let's consider a small perturbation of the coordinates $(U_1,\ldots,U_\nu,V_1,\ldots,V_\nu,W_0,\ldots,W_{2g-\nu})$. Starting with any solution U,V,W to the equation $f-V^2 = UW$ (with prescribed degrees) we will show that the vector space of triples $\dot U,\dot V,\dot W$ ($\deg \dot U,\dot V \le \nu-1$, $\deg \dot W \le 2g-\nu$) such that

$$f-(V+\varepsilon\dot V)^2 \equiv (U+\varepsilon\dot U)(W+\varepsilon\dot W) \bmod \varepsilon^2 \qquad (*)$$

has dimension ν.

The dimension must be $\ge \nu$ since in general k equations in n-dimensional affine space define a closed set whose irreducible components are varieties of dimension $\ge$ n-k; which in our case means $\ge (2g+1+\nu) - (2g+1) = \nu$.

On the other hand, the condition (*) is equivalent to the equation

$$(*) \qquad \dot{U}W + U\dot{W} + 2V\dot{V} = 0.$$

If we can prove that any polynomial of degree $\leq 2g$ can be written in the form $\dot{U}W + U\dot{W} + 2V\dot{V}$, then the number of linear conditions imposed by (*) equals the dimension of the space of polynomials in t of degree $\leq 2g$, which is $2g+1$, and we conclude that the dimension of the space of solutions of (*) equals $(2g+1+\nu)-(2g+1) = \nu$. But notice:

$W\dot{U}$ gives all the polynomials that are multiples of U

$2V\dot{V}$ assumes any given values at the points t where $U = 0$, $V \neq 0$

$\dot{U}W$ " " " " " " " " $U = 0$, $W \neq 0$

These make a vector space of dimension $(2g-\nu+1)+\nu = 2g+1$, since $U = V = W = 0$ never happens, or $f = V^2+UW$ would have a double zero.

2. $(C^\nu)_o \longrightarrow V(a_\alpha)$ is a morphism. First observe that the map is a morphism on the smaller Zariski-open set $(C^\nu)_{oo} = (C^\nu)_o - \bigcup_{i<j} p_{ij}^{-1}(\Delta)$, where $\Delta \subset C\times C$ is the diagonal. This is because the coefficients of U,V (hence W) were given above by an explicit formula as rational functions in the coordinates $s(P_i),t(P_i)$ with denominators products of $t(P_i)-t(P_j)$. These denominators are zero only if $P_i = P_j$ or $P_i = \iota(P_j)$ $(i \neq j)$, i.e., only on $p_{ij}^{-1}(\Delta \cup \Gamma)$.

To see that the map is a morphism elsewhere, we use Newton's Interpolation formula. This is expressed in

<u>Theorem 1.4</u> (Newton): <u>Let</u> f <u>be a</u> $\mathcal{C}^\infty$ <u>function on an open set</u> $U \subset \mathbb{R}$ <u>(resp. an analytic function on an open set</u> $U \subset \mathbb{C}$). <u>Define by induction on</u> n

$$f(x_1,\cdots,x_n) = \frac{f(x_1,\cdots,x_{n-1}) - f(x_2,\cdots,x_n)}{x_1 - x_n} .$$

<u>Then</u> f <u>is a</u> $\mathcal{C}^\infty$ <u>function (resp. an analytic function) on</u> U^n, <u>symmetric in its</u> n <u>arguments</u> x_i, <u>and for all</u> $a_1,\cdots,a_n$:

$$f(x) = \left[f(a_1)+(x-a_1)f(a_1,a_2)+\cdots+\prod_{i=1}^{n-1}(x-a_i)\cdot f(a_1,\cdots,a_n)\right]+\prod_{i=1}^{n}(x-a_i)\cdot f(x,a_1,\cdots,a_n).$$

Note that the expression in brackets is therefore the unique polynomial V(x) of degree $\leq$ n-1 such that:

$$\left(\frac{d}{dx}\right)^k [f(x)-V(x)]\Big|_{x=a} = 0, \qquad 0 \leq k \leq \begin{pmatrix}\#\text{ of } a_i\\ \text{equal to } a\end{pmatrix} - 1 .$$

To apply this to our problem, we define by induction on n rational functions on C^n by:

$$s(P_1,\cdots,P_n) = \frac{s(P_1,\cdots,P_{n-1})-s(P_2,\cdots,P_n)}{t(P_1)-t(P_n)} .$$

As in Newton's theorem, it is an easy calculation that $s(P_1,\cdots,P_n)$ is symmetric in $P_1,\cdots,P_n$. I claim

$$s(P_1,\cdots,P_n) \in \Gamma\left((C^n)_o, \mathcal{O}_{C^n}\right).$$

For $n = 2$, note that

$$s(P_1,P_2) = \frac{s(P_1)-s(P_2)}{t(P_1)-t(P_2)} = \frac{s^2(P_1)-s^2(P_2)}{[t(P_1)-t(P_2)][s(P_1)+s(P_2)]}$$

$$= \underbrace{\left[\frac{f(t(P_1))-f(t(P_2))}{t(P_1)-t(P_2)}\right]}_{\substack{\text{polynomial in}\\ t(P_1),t(P_2)}} \cdot \frac{1}{s(P_1)+s(P_2)}$$

Thus $s(P_1,P_2)$ has no poles in the open set $t(P_1) \neq t(P_2)$ nor in the open set $s(P_1) \neq -s(P_2)$ The union of these 2 open sets is $C^2-\Gamma$ since $t(P_1) = t(P_2)$ and $s(P_1) = -s(P_2)$ implies $P_2 = \iota(P_1)$. Thus $s(P_1,P_2)$ has no poles on $(C^2)_o$, hence is in $\Gamma((C^2)_o, \mathcal{O}_{C^2})$. For $n \geq 3$, by induction and the expression for $s(P_1,\cdots,P_n)$, $s(P_1,\cdots,P_n)$ has poles only if $t(P_1) = t(P_n)$. But by symmetry, it has poles only if $t(P_2) = t(P_n)$ too. The subset $t(P_1) = t(P_2) = t(P_n)$ has codimension 2 in $(C^n)_o$, so $s(P_1,\cdots,P_n)$ has no poles at all in $(C^n)_o$.

Finally, by Newton's theorem, the interpolating polynomial $V(t)$ can be expressed by:

$$V(t) = s(P_1)+(t-t(P_1)).s(P_1,P_2)+\cdots+\prod_{i=1}^{\nu-1}(t-t(P_i))\cdot s(P_1,\cdots,P_\nu) .$$

Thus the coefficients V_i of $V(t)$ are polynomials in $t(P_i)$ and $s(P_1,\cdots,P_k)$, hence are functions in $\Gamma((C^\nu)_o, \mathcal{O}_{C^\nu})$. This proves that $(C^\nu)_o \longrightarrow V(a_\alpha)$ is a morphism.

A consequence is that the set $V(a_\alpha)$ is irreducible since $(C^\nu)_o$ maps onto $V(a_\alpha)$ and $(C^\nu)_o$ is irreducible. To complete the proof, use the elementary:

Lemma 1.5: If $V \subset \mathbb{C}^n$ is an affine variety, $f_1,\cdots,f_k \in \mathbb{C}[X_1,\cdots,X_n]$ are polynomials such that

$$V = \left\{x \in \mathbb{C}^n \,\middle|\, f_i(x) = 0 \text{ all } i\right\}$$

$$\forall\, x \in V,\ T_{V,x} = \left\{\dot{x} \in \mathbb{C}^n \,\middle|\, f_i(x+\varepsilon\dot{x}) = 0 \bmod \varepsilon^2, \text{ all } i\right\},$$

then $(f_1,\cdots,f_k)$ is the prime ideal of all polynomials zero on V.

(Proof omitted).

We want to emphasize at this point the rather unorthodox use that we are making of the polynomials U,V,W:

a) we have a bijection

$$\begin{pmatrix}\text{divisors D on C}\\ \text{of a certain type}\end{pmatrix} \longleftrightarrow \begin{pmatrix}\text{three polynomials}\\ U(t),V(t),W(t) \text{ of a certain type}\end{pmatrix}$$

Thus

b) these divisors D become the points of a variety for which the coefficients of U,V,W are coordinates.

To take the coefficients of certain auxiliary polynomials as coordinates for a new variety is quite typical of moduli constructions, although it is usually not so explicitly carried out. In all of this Chapter, U,V,W will play the main role, and we will talk of (U,V,W) as representing a point of the variety $\mathrm{Div}_o^{+,\nu}(C)$.

Actually, for any smooth projective curve X, it's possible to describe $\mathrm{Div}^{+,\nu}(X)$ as a projective variety, although not as explicitly as in the above construction. We outline this without giving details, as it will not be used later. We use the bijection $\mathrm{Div}^{+,\nu}(X) \cong \mathrm{Symm}^{\nu}(X)$, the orbit space of X^{ν} under the action of the symmetric group permuting the factors.

A) Given an embedding $X \hookrightarrow \mathbb{P}^n$, we have the associated Segre embedding:

$$j: X^{\nu} \hookrightarrow \mathbb{P}^{(n+1)^{\nu}-1}$$

given by:

$$\forall\ P_{\alpha} = (X_0^{(\alpha)}, \cdots, X_n^{(\alpha)}) \in X$$

then $$(P_1, \ldots, P_{\nu}) \longmapsto (\cdots, \prod_{\alpha=1}^{\nu} X_{\sigma(\alpha)}^{(\alpha)}, \cdots)_{\sigma}$$

(one coordinate for every map $\sigma : \{1, \cdots, \nu\} \longrightarrow \{0, \cdots, n\}$).

B) j is equivariant under the action of the symmetric group on X^{ν}; on the homogeneous coordinate ring of X^{ν}, $R = \mathbb{C}[\cdots, Y_{\sigma}, \cdots]$, S_{ν} acts preserving the grading; the ring of invariants $R^{S_{\nu}}$ is finitely generated by homogeneous polynomials and $\exists M$ such that if $g_0, \cdots, g_N$ are a basis of $R^{S_{\nu}}$ in degree M, then:

$$R^{S_{\nu}} \supset \underset{\substack{\text{(elements w. degrees}\\ \text{divisible by M)}}}{R^{S_{\nu}}} = \mathbb{C}[g_0, \cdots, g_N],$$

C) Via g_i we have the embedding:

$$\begin{array}{ccc} \mathrm{Symm}^{\nu} X & \hookrightarrow & \mathbb{P}^N \\ \uparrow & & \uparrow \\ X^{\nu} & \hookrightarrow & \mathbb{P}^{(n+1)^{\nu}-1} \end{array}$$

D) The smoothness of $\mathrm{Symm}^{\nu} X$ follows from local analytic description:

$$\mathrm{Symm}^{\nu}(z\text{-disc.}) \underset{\text{biholomorphically}}{\cong} \{\text{open set in } \mathbb{C}^{\nu}\}$$

$$\text{via} \quad \sum_{i=1}^{\nu} P_i \longmapsto [\text{elem. symm. functions of } z(P_i)].$$

The explicit coordinates given by prop. 1.2 are particular to the case of hyperelliptic curves.

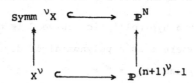

§2. Algebraic construction of the Jacobian of a hyperelliptic curve.

Let's recall that a hyperelliptic curve C is determined by an equation $s^2 = f(t)$, where f is a polynomial of degree 2g+1; C has one point at infinity, and $(t)_\infty = 2\cdot\infty$

$$(s)_\infty = (2g+1)\cdot\infty.$$

We shall study the structure of Pic C = {group of divisors modulo linear equivalence}.

Since the degree of the divisor (f) of a rational function is zero, there is a homomorphism

$$\deg\colon \text{Pic } C \longrightarrow \mathbb{Z}$$

$$\left(\text{divisor class } \sum n_i P_i\right) \longmapsto \sum n_i$$

Definition 2.1. The Jacobian variety of C is given by:

$$\text{Jac } C = \text{Ker}[\deg\colon \text{Pic } C \longrightarrow \mathbb{Z}]$$

We wish to endow Jac(C) with the structure of an algebraic variety. The possibility of doing this by purely algebraic constructions was discussed by A. Weil. In the hyperelliptic case, his construction becomes quite explicit. For the general case, see Serre [ ].

Step I. Given any g+1 points on the curve $P_1,\ldots,P_{g+1}$, such that $P_i \neq \infty$ and $P_i \neq \iota P_j$ if $i \neq j$, the function

$$\frac{s+\phi(t)}{\prod_{i=1}^{g+1}(t-t(P_i))} \qquad \text{where } \phi(t(P_i)) = s(P_i),\ \deg\phi \leq g,$$

has simple poles at all of the P_i's and no poles anywhere else, for numerator and denominator are both zero at ιP_i, and the numerator is $\neq 0$ at P_i [1]. At ∞, $(s)_\infty = (2g+1)\cdot\infty$, $(\phi(t))_\infty \leq 2g\cdot\infty$, $(\prod_{i=1}^{g+1}(t-t(P_i)))_\infty = (2g+2)\cdot\infty$, so the function is zero. Thus:

$$\sum_{i=1}^{g+1} P_i \equiv \infty + \sum_{i=1}^{g} Q_i, \text{ suitable } Q_i\text{'s}.$$

Consider also the function $\frac{t-a}{t-b}$ for any numbers a,b; it gives an equality of divisor classes:

$$P_a + \iota P_a = \left(\frac{t-a}{t-b}\right)_0 \equiv \left(\frac{t-a}{t-b}\right)_\infty = P_b + \iota P_b.$$

Similarly, using the function t-a, we see:

$$P_a + \iota P_a \equiv 2\infty .$$

Let's define $\mathcal{L}$ to be the divisor class of degree 2 that contains $P+\iota P$, all $P \in C$. [2]

The above remarks show that for every divisor D of degree zero $\exists P_1,\ldots,P_g$ such that $D \equiv \sum_{i=1}^{g} P_i - g\cdot\infty$. In fact for any D of degree 0, $D = \sum_{i=1}^{\ell} R_i - \sum_{i=1}^{\ell} S_i = \sum R_i + \sum \iota S_i - \sum(\iota S_i + S_i) \equiv \sum R_i + \sum \iota S_i - 2\ell\cdot\infty$; also write $2\cdot\infty$ whenever a pair $R+\iota R$ occurs in $\sum R_i + \sum \iota S_i$; now we can use the construction above in order to decrease the number of points

---

1) Or if P_i is a branch point, then the numerator vanishes to 1st order and the denominator to 2nd order at P_i.

2) Also called the fundamental "pencil" on the hyperelliptic curve; "pencil" because the projective dimension of $|P+\iota P| = \mathbb{P}(\mathcal{L}(P+\iota P))$ is 1. In the affine part of the curve $C_1 \subset \mathbb{C}^2$, $|P+\iota P|$ is cut out on C_1 by the pencil of lines through the point at infinity of the curve.

in $\sum R_i + \sum \iota S_i$ to $\leq g$. Therefore the map:

$$I: \ \mathrm{Symm}^g C \longrightarrow \mathrm{Jac}\ C$$
$$\sum P_i \longmapsto \sum P_i - g\cdot\infty$$

is surjective.

(This in fact is true for every curve.)

Step II. Given a divisor $\sum_{i=1}^{g} P_i$, $P_i \neq \infty$, $P_i \neq \iota P_j$ if $i \neq j$, then $\nexists$ a non constant rational function on C whose poles are bounded by $\sum P_i$.

Proof. Let h be such a function; then $h\cdot\prod_{i=1}^{g}(t-t(P_i))$ has poles only at ∞, hence is a polynomial in the affine coordinates s,t, i.e., it has the form $\phi(t)+s\psi(t)$, where ϕ and ψ are polynomials; now $v_\infty(s) = 2g+1$ which is odd, $v_\infty(\phi(t))$ is even, hence $v_\infty(s\psi) \neq v_\infty(\phi)$ so

$$0 = v_\infty(h) = v_\infty\left(\frac{\phi+s\psi}{\prod(t-t(P_i))}\right) \geq v_\infty(s\psi) - 2g = 1+v_\infty(\psi) \geq 1,$$

which is a contradiction unless $\psi(t) = 0$, i.e., h is a function of t only, $h\circ\iota = h$; this implies that the poles of h are bounded by $\sum \iota P_i$ also, thus h cannot have poles, i.e., is a constant.

Definition 2.2. Θ = [subset of Jac C of divisor classes of the form $\sum_{i=1}^{g-1} P_i-(g-1)\infty$].

Steps I and II imply that a suitable restriction of the map I is injective:

$$Z = \begin{pmatrix} \text{divisors } \sum_{i=1}^{g} P_i \text{ such that} \\ P_i \neq \infty, P_i \neq \iota P_j \text{ if } i \neq j \end{pmatrix} \xrightarrow[\text{res } I]{\cong} \text{Jac } C - \Theta$$

$$\bigcap \qquad\qquad\qquad\qquad \bigcap$$

$$\text{Symm}^g C \xrightarrow{\quad I \quad} \text{Jac } C \quad ;$$

in fact by Step II, $I(D) = I(D')$ for $D \in Z$ implies $D = D'$ because a function such that $D'-D = (h)$ would have poles only on $D = \sum_{i=1}^{g} P_i$, hence be a constant; in particular $Z \cap I^{-1}\Theta = \emptyset$ since Θ is the image of $\sum_{1}^{g-1} P_i + \infty$. Now if we represent any divisor class in Jac $C-\Theta$ as $\sum_{i=1}^{g} P_i - g\cdot\infty$ by Step I, then $\sum_{1}^{g} P_i$ is in Z, because if $P_i = \infty$ or $P_i = \iota P_j$, i.e., $P_i + P_j \equiv 2\cdot\infty$, then $(\sum_{1}^{g} P_i - g\cdot\infty) \in \Theta$.

By the previous section, Z is a smooth g-dimensional variety; by translation, we will cover Jac C by affine pieces isomorphic to Z.

Step III. Recall that $B \subset C$ is the set of branch points P, defined by $P = \iota P$: thus $2P = L$, for all $P \in B$.

Definition 2.3. Let $T \subset B$ be a subset of even cardinality; define

$$e_T = \Big(\sum_{P \in T} P\Big) - \Big(\frac{\#T}{2}\Big)\cdot L \in \text{Jac } C .$$

Lemma 2.4. a) $2e_T = 0$

b) $e_{T_1} + e_{T_2} = e_{T_1 \circ T_2}$ where $T_1 \circ T_2 = (T_1 \cup T_2) - (T_1 \cap T_2)$,

(symmetric difference)

c) $e_{T_1} = e_{T_2}$ if and only if $T_1 = T_2$ or $T_1 = CT_2$, the complement of T_2 in B.

Thus, the set of the e_T's forms a group isomorphic to $(\mathbb{Z}/2\mathbb{Z})^{2g}$.

Proof. a) $2e_T = \sum_{P \in T} 2P - (\#T)L \equiv 0$.

b) $e_{T_1} + e_{T_2} = \sum_{P \in T_1} P + \sum_{P \in T_2} P - \left(\frac{\#T_1 + \#T_2}{2}\right) \cdot L$; the P's that occur once in $\sum_{P \in T_1} P + \sum_{P \in T_2} P$ are those in $T_1 \circ T_2$; the others can be cancelled against L's because $2P \equiv L$, and the multiple k of L in $e_{T_1} + e_{T_2}$ is determined by $\deg\left(\sum_{P \in T_1 \circ T_2} P\right) = 2k$.

c) $e_T + e_{CT} = e_B = \sum_{P \in B} P - (g+1)L$: the function s has a simple zero at each of the branch points except ∞, and $(s)_\infty = (2g+1) \cdot \infty$, so $0 \equiv (s) = \left(\sum_{P \in B} P - \infty\right) - (2g+1)\infty = e_B$.

To prove the converse, it is enough to check that if $T \neq \emptyset$ or B then $e_T \neq 0$. By replacing if necessary T by CT, we may assume $\#T \leq g+1$ and, in the case $\#T = g+1$, ∞ is in T. $e_T = 0$ means $\sum_{P \in T} P \equiv \frac{\#T}{2} \cdot L \equiv \#T \cdot \infty$. Therefore there must be a function h with $(h)_\infty = \sum_{P \in T} P$; by putting ∞ on the right if it occurs, we bound the poles of h with at most g distinct branch points, none of which is ∞; since two distinct such P's cannot be conjugate, by step II f must be a constant and $T = \emptyset$.

<u>Lemma 2.5.</u>* $\bigcup_T \left[(\text{Jac } C - \Theta) + e_T\right] = \text{Jac } C$

<u>or</u> $\bigcap_T (\Theta + e_T) = \phi$.

<u>Proof</u>. Write any $D = \sum_{i=1}^{g} P_i - g\cdot\infty$ as $= \sum_{i=1}^{r} Q_i - r\infty$ with $Q_i \neq \infty$, $Q_i \neq \iota Q_j$ (by replacing $P + \iota P$ with 2∞ if it occurs). Now choose g-r branch points $R_1 \cdots R_{g-r}$ distinct from the Q_i's and ∞ . Then

$$D + \left(\sum_{i=1}^{g-r} R_i - (g-r)\cdot\infty\right) = \sum Q_i + \sum R_i - g\infty \in \left[\text{Jac } C - \Theta\right]$$

(because it's the image of a point in Z). If g-r is even, then $D \in (\text{Jac } C-\Theta) + e_{\{R_1,\cdots,R_{g-r}\}}$; if g-r is odd

$D \in (\text{Jac } C-\Theta) + e_{\{R_1,\ldots,R_{g-r},\infty\}}$. <u>QED</u>

So, we take one copy of Z for each T, and we glue them together according to their identification as subsets of the Jacobian; we have to see that this glueing satisfies the conditions to give the atlas of a variety.

---

*Here - in Jac C-Θ is a difference of sets, but + in $(\text{Jac } C-\Theta)+e_T$ means translation of a set by a point using the group law on Jac C.

Step IV.

Lemma 2.6. *Given any* $\{B_i\}_{i\in T_1}$, $\{B_i\}_{i\in T_2} \subset B$, *with* T_1, T_2 *of even cardinality, let* $\Gamma_{T_1,T_2} \subset Z\times Z$ *be the set of pairs*

$$\left\{ \sum_{i=1}^{g} P_i, \sum_{i=1}^{g} Q_i \;\middle|\; \sum_{i=1}^{g} P_i + e_{T_1} \equiv \sum_{i=1}^{g} Q_i + e_{T_2} \right\}.$$

Then Γ_{T_1,T_2} *is Zariski closed and projects isomorphically to Zariski open subsets of each factor.*

Proof. Rewrite the definition of Γ_{T_1,T_2} as

$$\left\{ \sum P_i, \sum Q_i \;\middle|\; \sum_{i=1}^{g} P_i + \sum_{i=1}^{g} \iota Q_i + \sum_{i\in T_1} B_i + \sum_{i\in T_2} B_i \equiv (2g+\#T_1+\#T_2)\cdot\infty \right\}.$$

Consider the vector space V of functions whose poles are bounded by $N\cdot\infty$ where $N = 2g+\#T_1+\#T_2$; as we saw before (lemma 1.1)

$$V = \begin{pmatrix}\text{polynomials in } t \\ \text{of degree} \le \left[\frac{N}{2}\right]\end{pmatrix} + s\begin{pmatrix}\text{polynomials in } t \\ \text{of degree} \le \left[\frac{N-2g-1}{2}\right]\end{pmatrix}.$$

Say $f_1,\dots,f_M$ is a basis of V, where $M = N-g+1$.

Among these functions, those which have zeroes at

$\sum_{i=1}^{g} P_i + \sum_{i=1}^{g} \iota Q_i + \sum_{i\in T_1} B_i + \sum_{i\in T_2} B_i$ for fixed $\sum P_i$, $\sum Q_i$, are just[3)] the elements of the ideal in $\mathbb{C}[s,t]/(s-f(t))$ given by the product:

---

3) Use the fact that $\mathbb{C}[s,t]/(s^2-f(t))$ is a Dedekind domain (§1, Note 1) plus Corollary 0.14.

$$(*)\quad I_{\Sigma P_i,\Sigma Q_i} = \left(U^{(1)}(t), s-V^{(1)}(t)\right)\cdot\left(U^{(2)}(t), s+V^{(2)}(t)\right)\cdot$$

$$\cdot \prod_{i\in T_1}(s,t-a_i)\cdot\prod_{i\in T_2}(s,t-a_i)$$

where the divisor $\sum_{i=1}^{g} P_i \longleftrightarrow (U^{(1)},V^{(1)},W^{(1)})$

and $\sum_{i=1}^{g} Q_i \longleftrightarrow (U^{(2)},V^{(2)},W^{(2)})$.

Note that if $h \in V \cap I_{\Sigma P_i,\Sigma Q_i}$ then $(\sum P_i, \sum Q_i) \in \Gamma_{T_1,T_2}$ since h has exactly N zeroes, and poles only at ∞.

Note also that membership in I imposes N linear conditions on a function, so $\operatorname{codim} I = 2g + \#T_1 + \#T_2 = N$, independent of $\sum P_i, \sum Q_i$.

Let $R_Z = (\text{affine ring of } Z) = \mathbb{C}[U_i,V_j,W_k]/(a_\alpha)$. Then we get a "universal" I

$$I \subset R_Z^{(1)} \otimes R_Z^{(2)}[s,t]/(s^2-f(t))$$

defined by the same formula (*) with $U_i^{(1)},V_i^{(1)} \in R_Z^{(1)}$, $U_i^{(2)},V_i^{(2)} \in R_Z^{(2)}$ being variables. Consider

$$A = R_Z^{(1)} \otimes R_Z^{(2)}[s,t]\big/(s^2-f(t))+I;$$

A is an algebra over $R_Z^{(1)} \otimes R_Z^{(2)}$, finitely generated and integrally dependent (I contains a monic polynomial in t) such that for all homomorphisms $R_Z^{(1)} \otimes R_Z^{(2)} \longrightarrow \mathbb{C}$ (evaluation of the coordinates) it becomes a $\mathbb{C}$-vector space of fixed dimension N. It follows[4] that A is "locally free", i.e.,

$$\exists h_\alpha, g_\alpha \in R_Z^{(1)} \otimes R_Z^{(2)} \quad \text{such that } 1 = \sum h_\alpha g_\alpha$$

and $\forall\alpha\ \exists\ e_1^{(\alpha)}, \cdots, e_N^{(\alpha)}$ basis of A_{h_α} as $(R_Z^{(1)} \otimes R_Z^{(2)})_{h_\alpha}$-module.

Now let the map

$$V \longrightarrow A$$

be given by
$$f_i \longmapsto \sum c_{ij}^{(\alpha)} e_j^{(\alpha)} .$$

"$\mathrm{rk}(c_{ij}^{(\alpha)}) < N-g+1$" defines Γ_{T_1,T_2} in the open set $h_\alpha \neq 0$ of $Z \times Z$

---

4) This follows from the

Proposition. If R is the affine ring of an affine variety, S a finitely generated R-module, and

$$\dim_{\mathbb{C}} S \otimes_R R/\mathfrak{m}$$

is constant as $\mathfrak{m}$ varies among the maximal ideals of R, then S is a locally free R-module.

Proof: If $\mathfrak{m}$ is a maximal ideal in R, let $e_1, \cdots, e_N \in S_{\mathfrak{m}}$ be a basis for the vector space $S_{\mathfrak{m}} \otimes_{R_{\mathfrak{m}}} R_{\mathfrak{m}}/\mathfrak{m}R_{\mathfrak{m}} \cong S \otimes_R R/\mathfrak{m}$; by Nakayama's lemma, $e_1, \cdots, e_N$ generate $S_{\mathfrak{m}}$ as $R_{\mathfrak{m}}$-module; we claim they are a free set of generators. Since S is finite we can express its generators as combinations $\sum(\frac{g_i}{f})e_i$, $\frac{g_i}{f} \in R_{\mathfrak{m}}$, involving only one denominator $f \notin \mathfrak{m}$; it follows that $e_1, \cdots, e_N$ generate $R_{\mathfrak{n}}$, for any max. ideal $\mathfrak{n}$ of R which doesn't contain f. Now if there were a relation among the $e_1, \cdots, e_N$, $\sum \lambda_i e_i = 0$, $\lambda_i \in R_{\mathfrak{m}}$ and say $\lambda_1 \neq 0$, let's express the λ_i's as g_i'/f', $g_i', f' \in R$. Then $g_i' \in \mathfrak{m}$, since $e_1, \ldots, e_N$ is a basis for $S_{\mathfrak{m}}/\mathfrak{m}S_{\mathfrak{m}}$. There is a maximal ideal $\mathfrak{n}$ such that $g_1 ff' \notin \mathfrak{n}$, since R is the ring of an affine variety. But if $g_1 ff' \notin \mathfrak{n}$ then λ_1 is not zero in $R/\mathfrak{n}R$ and $\dim_{\mathbb{C}} S_{\mathfrak{n}} \otimes_{R_{\mathfrak{n}}} R/\mathfrak{n}R < N$, which contradicts the assumption. (See Hartshorne, Algebraic Geometry, Ex. 5.8, p. 125.)

(such open sets cover $Z \times Z$ because $\sum h_\alpha g_\alpha = 1$). This proves that Γ_{T_1,T_2} is Zariski-closed.

Note: the rank of $(c_{ij}^{(\alpha)})$ is never less than N-g since 2 functions in the kernel of $V \longrightarrow A$ must be linearly dependent, having the same zeroes and poles. Therefore Γ is defined locally by g equations. This follows from:

Proposition 2.7. Given a matrix

$$M = \begin{pmatrix} a_{11}(x) & \cdots & a_{1n}(x) & a_{1,n+1}(x) & \cdots & a_{1m}(x) \\ \vdots & & \vdots & \vdots & & \vdots \\ a_{n1}(x) & \cdots & a_{nn}(x) & a_{n,n+1}(x) & \cdots & a_{nm}(x) \end{pmatrix}$$

where $a_{ij}(x)$ are polynomial functions on an affine variety, let

$$M_{I,J} = \det_{\substack{i \in I \\ j \in J}}(a_{ij}), \quad \#I = \#J,$$

and suppose $M_{(1,\ldots,n-1),(1,\ldots,n-1)}(0) \neq 0$; then $\exists$ a Zariski open neighborhood U of 0 such that for all $x \in U$, "$M_{(1,\ldots,n),(1,\ldots,n-1,i)}(x) = 0$ for $n \leq i \leq m$" if and only if "rk $M(x) = n-1$".

Therefore all components of Γ_{T_1,T_2} have dimension $\geq$ g. But each projection $\Gamma_{T_1,T_2} \longrightarrow Z$ is injective and dim Z = g. Therefore Γ_{T_1,T_2} must be irreducible and of dimension g too. By Zariski's Main Theorem (0.6), Γ_{T_1,T_2} is isomorphic to an open subset of Z under each projection.

This proves that Γ_{T_1,T_2} can be used to glue the T_1 and T_2 copies of Z. This procedure therefore constructs Jac C as a variety; we will see later that it is projective. Note that it is complete, because there is a surjective map $C^g \longrightarrow$ Jac C, and C^g is compact in the complex topology as C is complete.

In fact, Jac C is an abelian variety:

Definition 2.8. An abelian variety X is a complete variety with a commutative group law such that addition $X \times X \longrightarrow X$ and inverse $X \longrightarrow X$ are morphisms.

We know that Jac(C) is a complete variety and a commutative group. To see that the group law is a morphism we need:

Lemma 2.9. For all T_1, T_2, T_3 even sets of branch points

$$\{\textstyle\sum P_i, \sum Q_i, \sum R_i \mid \sum_1^g P_i + \sum_1^g Q_i + \sum_{T_1} B_i + \sum_{T_2} B_i \equiv \sum_{i=1}^g R_i + \sum_{T_3} B_i + (g + \#T_1 + \#T_2 - \#T_3)\infty\}$$

is Zariski closed in $Z \times Z \times Z$, and projects isomorphically via p_{12} to $Z \times Z$.

(This can be proved in the same way as Lemma 2.6, so we omit the details.)

Proposition 2.10. As a complex manifold, every abelian variety X of dimension n is a complex torus $\mathbb{C}^n/L$.

Proof: We use the Lie group structure of X. The exponential mapping $\mathbb{C}^n$ = (Lie algebra of X) $\longrightarrow$ X is a homomorphism because X is commutative thus, being a diffeomorphism in a neighborhood of the identity, it is open and since the image is connected exp is surjective. Again by bijectivity in a neighborhood of 0, the kernel is a discrete subgroup of $\mathbb{C}^n$, and the only discrete subgroups L such that $\mathbb{C}^n/L$ is compact are lattices.

QED.

In fact, we already showed in Chapter II that Jac C was a complex torus (Abel's theorem II.2.5). We have thus a 2nd proof of this based on the chain of reasoning:

Jac C is a complete variety ~~~~> Jac C is an abelian variety ~~~~> Jac C is a complex torus.

<u>Corollary 2.11</u>. <u>Every</u> 2-<u>torsion element of</u> Jac C <u>is of the form</u> e_T, <u>some</u> $T \subset B$.

<u>Proof</u>: The 2-torsion subgroup of the abstract group $\mathbb{C}^g/L = \mathbb{R}^{2g}/\mathbb{Z}^{2g}$ is $\frac{1}{2}L/L$, which is isomorphic to $(\mathbb{Z}/\mathbb{Z})^{2g}$. But it contains the group of the e_T's, which has the same order, so they coincide. QED.

§3. The translation-invariant vector fields

Let X be a variety. Then a vector field D on X is given equivalently by:

a) a family of tangent vectors $D(x) \in T_{X,x}$, all $x \in X$ such that in local charts

$$X \supset U_\alpha \subset \mathbb{C}^{n_\alpha}$$

$$D(x) = \sum_{i=1}^{n_\alpha} a_i(x).\,\partial/\partial x_i, \quad a_i \in \mathbb{C}[X_1,\cdots,X_{n_\alpha}].$$

b) a derivation $D: \mathcal{O}_X \longrightarrow \mathcal{O}_X$.

In fact, given $D(x)$, $f \in \Gamma(U,\mathcal{O}_X)$, define Df by

$$Df(x) = D(x)(f) .$$

When X is an abelian variety, then translations on X define isomorphisms

$$T_{X,0} \xrightarrow{\sim} T_{X,x}$$

for all $x \in X$ (0 = identity), so we may speak of translation-invariant vector fields. It is easy to see that for all $D(0) \in T_{X,0}$, there is a unique translation-invariant vector field with this value at 0. In general, the vector fields on X form a Lie algebra under commutators:

$$[D_1,D_2](f) = D_1D_2f-D_2D_1f.$$

For translation-invariant vector fields, the commutativity of X implies that bracket is zero (see Abelian Varieties, D. Mumford, Oxford Univ. Press, p. 100.

The purpose of this section is to give explicit formulas for the invariant vector fields in the chart Z in Jac C. Our method is this. Let $P \in C$ and choose a non-zero $\delta_P \in T_{C,P}$. Let $\varepsilon \longmapsto P(\varepsilon) \in C$ be analytic coordinates in a small neighborhood of P with $P(0) = P$, so that δ_P is the image of the unit tangent vector $\partial/\partial\varepsilon$ at 0 in this coordinate. Then we get

$$D_P(0) \in T_{\text{Jac } C,0}$$

defined as the image of $\partial/\partial\varepsilon$ for the map

$$\begin{aligned} \varepsilon\text{-disc} &\longrightarrow \text{Jac } C \\ \varepsilon &\longmapsto [\text{divisor class } P(0)-P(\varepsilon)] \end{aligned}$$

at $\varepsilon = 0$, i.e., the tangent vector to this little analytic curve in Jac C at 0. Note that δ_P and $D_P(0)$ are determined by P only up to a scalar.

Starting with any divisor $D = \left(\sum_{i=1}^{g} P_i - g\cdot\infty\right) \in Z$, let

$$D - P(\varepsilon) + P \equiv \sum_{i=1}^{g} P_i(\varepsilon) - g\cdot\infty$$

and let

$$\sum_{i=1}^{g} P_i(\varepsilon) \longleftrightarrow (U_\varepsilon(t), V_\varepsilon(t), W_\varepsilon(t)).$$

Since Z is open, choosing $|\varepsilon|$ small enough, we can suppose $(D - P(\varepsilon) + P) \in Z$.

Then

$$\left(\frac{dU_\varepsilon}{d\varepsilon}\bigg|_{\varepsilon=0}, \frac{dV_\varepsilon}{d\varepsilon}\bigg|_{\varepsilon=0}, \frac{dW_\varepsilon}{d\varepsilon}\bigg|_{\varepsilon=0}\right) \in T_{Z,D} = T_{\text{Jac } C,D}$$

and this represents the translate of $D_P(0)$ to $T_{\text{Jac } C,D}$. Note that for this to be an invariant vector field, it is possible to use different uniformizations $\varepsilon \longmapsto P(\varepsilon)$ for each D, so long as the tangent vector δ_P to this map is independent of D.

The result is:

Theorem 3.1. For any $P \in C$, $P \neq \infty$, for suitable δ_P the above tangent vector is given at $(U,V,W) \in Z$ by

$$\dot{U}(t) = \frac{V(t(P))\cdot U(t) - U(t(P))\cdot V(t)}{t - t(P)}$$

$$\dot{V}(t) = \frac{1}{2}\,\frac{U(t(P))\cdot W(t) - W(t(P))\cdot U(t)}{t - t(P)} - U(t(P))\cdot U(t)$$

$$\dot{W}(t) = \frac{W(t(P))V(t) - V(t(P))\cdot W(t)}{t - t(P)} + U(t(P))\cdot V(t) \ .$$

Note. Equivalently, this means we have a derivation $D_P\colon \mathbb{C}[U_i,V_j,W_k]/(a_\alpha) \circlearrowleft$ given by

$$D_P(U_i) = \left[\text{coeff. of } t^{g-i} \text{ in } \frac{V(t(P))\cdot U(t) - U(t(P))\cdot V(t)}{t - t(P)}\right]$$

$D_P(V_i)$, $D_P(W_i)$ = [coeff. of t^{g-i} in the other expressions] .

Note. Corresponding to $P = \infty$, we get the vector field

$$\dot{U}(t) = V(t)$$

$$\dot{V}(t) = \frac{1}{2}[-W(t) + (t-U_1+W_0)U(t)]$$

$$\dot{W}(t) = -(t-U_1+W_0)\cdot V(t),$$

obtained by letting $t(P)$ go to ∞, and replacing δ_P in the Theorem by $\delta_P/t(P)^{g-1}$. To check this, we calculate:

$$\lim_{P\to\infty} \frac{\dot{U}(t)}{t(P)^{g-1}} = \lim\left[\frac{-t(P)^g\cdot V(t)+\text{lower order terms in } t(P)}{(-t(P)+t)\cdot t(P)^{g-1}}\right] = V(t),$$

$$\lim_{P\to\infty} \frac{\dot{V}(t)}{t(P)^{g-1}} = \lim \frac{1}{2}\left[\frac{t(P)^g W(t)-(t(P)^{g+1}+W_0 t(P)^g)U(t)+\begin{pmatrix}\text{lower terms}\\ \text{in } t(P)\end{pmatrix}-(t(P)^g+U_1 t(P)^{g-1})(t-t(P))U(t)}{(-t(P)+t)t(P)^{g-1}}\right.$$

$$= \lim \frac{1}{2}\left[\frac{t(P)^g W(t)-W_0 t(P)^g U(t)+U_1 t(P)^g U(t)-t(P)^g tU(t)+\begin{pmatrix}\text{lower terms}\\ \text{in } t(P)\end{pmatrix}}{-t(P)^g + \text{lower order terms in } t(P)}\right]$$

$$= \frac{1}{2}\left[-W(t) + (t-U_1+W_0)\cdot U(t)\right],$$

$$\lim_{P\to\infty} \frac{\dot{W}(t)}{t(P)^{g-1}} = \lim \frac{(t(P)^{g+1}+W_0 t(P)^g)V(t)+(t-t(P))(t(P)^g+U_1 t(P)^{g-1})V(t)+\begin{pmatrix}\text{lower terms}\\ \text{in } t(P)\end{pmatrix}}{(-t(P)+t)t(P)^{g-1}}$$

$$= -(t-U_1+W_0).V(t).$$

Note. $D_P(O) \in T_{\text{Jac } C,O}$ will depend on P and on the chosen uniformization; as P varies, we should only have g independent vector fields. To see this, it suffices to expand the above expressions in powers of $t(P)$. As before let:

$$U(t) = \sum_{i=0}^{g} U_i \, t^{g-i}, \qquad U_0 = 1$$

$$V(t) = \sum_{i=0}^{g} V_i \, t^{g-i}, \qquad V_0 = 0$$

$$W(t) = \sum_{i=-1}^{g} W_i \, t^{g-i}, \qquad W_{-1} = 1;$$

then

$$\dot{U}(t) = \sum_{i,j=0}^{g} V_i U_j \frac{t(P)^{g-i} t^{g-j} - t(P)^{g-j} t^{g-i}}{t - t(P)}$$

$$= \sum_{\substack{0 \le i < j \le g \\ \text{so } g-i > g-j}} V_i U_j t(P)^{g-j} t^{g-j} \frac{t(P)^{j-i} - t^{j-i}}{t - t(P)} + \sum_{\substack{g \ge i > j \ge 0 \\ \text{so } g-i < g-j}} V_i U_j t(P)^{g-i} t^{g-i} \frac{t^{i-j} - t(P)^{i-j}}{t - t(P)}$$

$$= \sum_{i>j} (V_i U_j - V_j U_i) \underbrace{t(P)^{g-i} t^{g-i} \left(t^{i-j-1} + \cdots + t(P)^{i-j-1}\right)}_{\sum\limits_{\substack{k+\ell = i+j+1 \\ 1 \le j+1 \le k, \ell \le i \le g}} t(P)^{g-k} t^{g-\ell}}$$

$$= \sum_{k=1}^{g} t(P)^{g-k} \left[\sum_{\ell=1}^{g} t^{g-\ell} \sum_{\substack{i+j = (k+\ell)-1 \\ i \ge \max(k,\ell) \\ j \le \min(k,\ell)-1}} (V_i U_j - V_j U_i) \right]$$

$$2\dot{V}(t) = \sum_{\substack{0<i<g\\ -1\leq j\leq g}} U_i W_j \frac{t(P)^{g-i}t^{g-j} - t(P)^{g-j}t^{g-i}}{t-t(P)} - \sum_{0\leq i,j\leq g} U_i U_j t(P)^{g-i}t^{g-j}$$

$$= \sum_{i>j} (U_iW_j - U_jW_i)\, t(P)^{g-i}t^{g-i}(t^{i-j-1}+\dots+t(P)^{i-j-1}) - \sum U_iU_j t(P)^{g-i}t^{g-j}$$

$$= \sum_{k=1}^{g} t(P)^{g-k}\left[\sum_{\ell=1}^{g} t^{g-\ell}\left(\sum_{\substack{i+j=k+\ell-1\\ i\geq \max(k,\ell)\\ j\leq \min(k,\ell)-1}} (U_iW_j - U_jW_i) - U_kU_\ell\right)\right]$$

(k and ℓ are allowed to run from 0 to g, but for $k = 0, \ell = 0$
$\sum(U_iW_j - U_jW_i) - U_kU_\ell = U_\ell W_{-1} - U_\ell, = U_kW_{-1} - U_k$ (resp.) $= 0$).

$$\dot{W}(t) = \sum W_iV_j \frac{t(P)^{g-i}t^{g-j} - t(P)^{g-j}t^{g-i}}{t - t(P)} + \sum U_iV_j t(P)^{g-i}t^{g-j}$$

$$= \sum_{k=1}^{g} t(P)^{g-k}\left[\sum_{\ell=0}^{g} t^{g-\ell}\left(\sum_{\substack{i+j=k+\ell-1\\ i\geq \max(k,\ell)\\ j\leq \min(k,\ell)-1}} (W_iV_j - W_jV_i) + U_kV_\ell\right)\right]$$

(k is allowed to be zero but for $k = 0$, $\sum(W_iV_j - W_jV_i) + U_kV_\ell = 0$,
while for $\ell = 0$ we get $-\sum_{k=1}^{g} t(P)^{g-k}t^gV_k$).

So if

$$D_k U_\ell = \sum_{\substack{i+j=k+\ell-1 \\ i>\max(k,\ell) \\ j\le\min(k,\ell)-1}} (V_i U_j - V_j U_i)$$

$$D_k V_\ell = \frac{1}{2}\Big[\sum_{\substack{\text{same} \\ \text{as above}}} (U_i W_j - U_j W_i) - U_k U_\ell\Big]$$

$$D_k W_\ell = \sum_{\substack{\text{same} \\ \text{as above}}} (W_i V_j - W_j V_i) + U_k V_\ell$$

then we find $D_P = \sum_{k=1}^{g} t(P)^{g-k} D_k$.

Proof of the theorem 3.1. For the proof, we also assume $P \notin \operatorname{Supp} \sum_{i=1}^{g} P_i$, and that neither P nor any P_i is a branch point. The result will follow by continuity for all P and $\sum P_i$. Let $\sum_{i=1}^{g} P_i$ correspond to (U,V,W) as usual and note that as no P_i is a branch point, U,V have no common zeroes.

We consider the function

$$q(s,t) = \frac{U(P).(s+V(t))+U(t).(s(P)-V(P))}{U(t).(t-t(P))} ;$$

the denominator is zero at $\sum_1^g P_i + \sum_1^g \iota(P_i)+P+\iota(P)$, but the numerator is zero at $\sum_1^g \iota(P_i)+\iota(P)$ so q has poles at $\sum_{i=1}^{g} P_i$ and at P. Its principal part at P is $\frac{2s(P)}{t-t(P)}$, independent of $\sum P_i$. At infinity, q is like $U(P)\cdot\frac{s}{t^{g+1}}$ so q has a zero at ∞.

So the equation $q(s,t)^{-1} = \frac{\varepsilon}{2}$

$$\text{or} \quad U(t)(t-t(P)) = \frac{\varepsilon}{2}[U(P)(s+V(t))+U(t)(s(P)-V(P))]$$

$$\text{has solutions} \begin{cases} \sum_{i=1}^{g} P_i(\varepsilon) & \text{near} \quad \sum_{i=1}^{g} P_i \\ P(\varepsilon) & \text{near} \quad P, \end{cases}$$

$$\text{and} \quad \left[\sum_1^g P_i(\varepsilon)+P(\varepsilon)\right] - \left[\sum_1^g P_i+P\right] = \left(q - \frac{2}{\varepsilon}\right) \equiv 0$$

Unfortunately, this analytic family $\varepsilon \longmapsto P(\varepsilon)$ of points near P depends on the choice of U,V,W, hence on the divisor $\sum P_i$. But since the principal part of q at P is independent of $\sum P_i$, the tangent vector δ_P to the family $\varepsilon \longmapsto P(\varepsilon)$ at $\varepsilon = 0$ is independent of $\sum P_i$. In fact, this gives

$$\frac{\varepsilon}{2}U(P)s = U(t).(t-t(P)) - \frac{\varepsilon}{2}[U(P)V(t)-U(t)V(P)+U(t)s(P)] \;;$$

squaring both sides

$$\frac{\varepsilon^2}{4} U(P)^2 f(t) = U(t)^2[t-t(P)]^2 - U(t).(t-t(P)).\varepsilon[U(P)V(t)-U(t)V(P) +$$

$$+ U(t)s(P)] + \frac{\varepsilon^2}{4}[U(P)^2V(t)^2+2U(P)V(t)U(t)(s(P)-V(P))+U(t)^2(s(P)-V(P))^2]$$

or (substituting $f(t) = V(t)^2+U(t)W(t)$ and dividing by $U(t)(t-t(P))$)

$$U(t).(t-t(P))-\varepsilon[U(P)V(t)-U(t)V(P)+U(t)s(P)] +$$

$$+ \frac{\varepsilon^2}{4}\left[\frac{-U(P)^2W(t)+2U(P)V(t)(s(P)-V(P))+U(t)(s(P)-V(P))^2}{t - t(P)}\right] = 0$$

$$\text{or}\quad 0 = \underbrace{\left[U(t) + \varepsilon\,\frac{U(t)V(P)-U(P)V(t)}{t-t(P)} + \varepsilon^2(\quad)+\dots\right]}_{\text{degree } g \text{ in } t}\underbrace{\left[t-t(P)-\varepsilon\, s(P)+\varepsilon^2(\quad)+\dots\right]}_{\substack{\text{degree 1 in } t;\\ \text{defines } P(\varepsilon)}}$$

Thus the 1st factor is $U_\varepsilon(t)$, hence differentiating, we find:

$$\dot{U}(t) = \frac{U(t)V(P)-U(P)V(t)}{t-t(P)} .$$

Now from the relation

$$f - (V+\varepsilon\dot{V})^2 \equiv (U+\varepsilon\dot{U})(W+\varepsilon\dot{W}) \quad \text{mod. } \varepsilon^2, \quad \text{i.e.,}$$

$$\begin{aligned} -2V\dot{V} &= \dot{U}W + \dot{W}U \\ &= \frac{U\,V(P) - U(P)\,V}{t - t(P)}\,W + \dot{W}U \end{aligned}$$

we have

$$0 = V\left(2\dot{V} - \frac{U(P)W}{t-t(P)}\right) + U\left(\dot{W} + \frac{V(P)W}{t-t(P)}\right)$$

Therefore

$$\begin{aligned} 2\dot{V} - \frac{U(P)W}{t-t(P)} &= -\,a(t)U \\ \dot{W} + \frac{V(P)W}{t-t(P)} &= +\,a(t)V \qquad \text{where } a \text{ is a rational function,} \end{aligned}$$

or

$$\dot{V} = \frac{1}{2}\left[\frac{U(P)\cdot W}{t-t(P)} - a(t)U\right]$$

$$\dot{W} = a(t)\cdot V - \frac{V(P)W}{t-t(P)}$$

Set $a(t) = \frac{W(P)}{t-t(P)} + \tilde{a}(t)$. Then

$$\dot{V} = \frac{1}{2}\left[\frac{U(P)W-W(P)U}{t-t(P)} - \tilde{a}(t)U\right]$$

$$\dot{W} = \frac{W(P)V-V(P)W}{t-t(P)} + \tilde{a}(t)V$$

so $\tilde{a}U$, $\tilde{a}V$ are polynomials in t; it follows that $\tilde{a}$ is a polynomial (since U,V are relatively prime), and since $\deg \dot{V} < g$ and $\dot{V} = [U(P)t^g - \tilde{a}(t)t^g + (\text{lower order terms in } t)]$, then $\tilde{a}(t) = U(P)$.

If U,V have common zeroes, the formula follows by continuity.

<u>QED</u>

In fact, we have something more here than an expression for the invariant vector fields on Jac C. Suppose we let the curve $s^2 = f(t)$ vary too. We see that we have a morphism

$$\mathbb{C}^{3g+1} \cong \left[\begin{array}{l}\text{space of } \underline{\text{all}} \text{ polynomials } U,V,W\\ \text{s.t. } U \text{ monic, deg. } g\\ \qquad V \qquad\quad \text{deg.} \leq g-1\\ \qquad W \text{ monic, deg. } g+1\end{array}\right], \quad \text{coord. } U_i, V_j, W_k$$

$$\downarrow \pi$$

$$\mathbb{C}^{2g+1} \cong \left[\begin{array}{l}\text{space of polynomials } f\\ \text{s.t. } f \text{ monic, deg } 2g+1\end{array}\right], \quad \begin{array}{c}\text{coord. } a_\alpha\\ f(t) = t^{2g+1} + a_1 t^{2g} + \cdots + a_{2g+1}\end{array}$$

where π is defined by:

$$f = V^2 + UW.$$

The fibre of π over any f with distinct roots is the affine piece Z of the Jacobian of $s^2 = f(t)$. Thus all Z's fit together into a fibre system. The formulae above define vector fields D_k, $1 \leq k \leq g$, on all of $\mathbb{C}^{3g+1}$, which are tangent to the subvarieties $\pi^{-1}(f)$ (and generate their tangent spaces at each point). Thus

$$[D_{k_1}, D_{k_2}] = 0$$

$$D_k(a_\alpha) = 0$$

(because the Jacobians are commutative groups, hence $[D_{k_1}, D_{k_2}]$ is zero on each Z)

To summarize we have found <u>an explicit set of g commuting vector fields on</u> $\mathbb{C}^{3g+1}$, <u>with</u> 2g+1 <u>polynomial invariants</u> a_α, <u>and integral manifolds</u> Jac C - Θ, <u>where</u> C <u>varies over all hyperelliptic curves.</u>

§4. Neumann's dynamical system.

In classical mechanics, one encounters the class of problems:

M = real 2n-dimensional manifold, with a closed non-degenerate differential 2-form ω

$\hat{\omega}$ = dual skew-symmetric form on T_M^*

H = $\mathcal{C}^\infty$-function on M, called the Hamiltonian.

$$X_H = \begin{cases} \text{the vector field on M defined by} \\ \omega(X_H,Y) = \langle Y,dH\rangle \text{ for all vectors } Y \\ \text{or} \\ \hat{\omega}(dH,\alpha) = \langle X_H,\alpha\rangle \text{ for all 1-forms } \alpha \end{cases}$$

Recall that we define the Poisson bracket by:

$$\{f,g\} = +\langle X_f,dg\rangle = -\langle X_g,df\rangle = \hat{\omega}(df,dg),$$

and that the compatibility condition

$$[X_f,X_g] = X_{\{f,g\}}$$

holds.

Moreover, $\{f,g\} = 0$ means that dg is perpendicular to X_f, or that g is constant on the orbits of the integral flow of X_f. The main problems of classical mechanics were all to integrate various vector fields X_H.

Unfortunately, it never happens except in trivial cases that there exist 2n-1 functions defining $M \xrightarrow{\pi} \mathbb{R}^{2n-1}$ such that $\pi^{-1}(x)$ = orbits of X_H. However, what does happen occasionally and to date unpredictably is that there exists a C^∞-map $h: M \longrightarrow U$, U open in $\mathbb{R}^n$. (a) if X_i are the coordinates on $\mathbb{R}^n$, $\{X_j \circ h, X_j \circ h\} = 0$, (b) h is submersive, (c) h is proper, and (d) $H = f \circ h$, f a C^∞ function on U. In this case, if $H_i = X_i \circ h$, it follows from $\{H_i, H_j\} = 0$ that the X_{H_i} commute, hence the fibres of h are n-dimensional compact submanifolds whose connected components are orbits of $\{X_{H_1}, \cdots, X_{H_n}\}$ hence are isomorphic to real tori: $(\mathbb{R}^n/\text{lattice})$. It can be proven that near each one of these tori M has coordinates $x_1, \cdots, x_n, y_1, \cdots, y_n$, x_i determined mod $\mathbb{Z}$, $H_i = H_i(y_1, \cdots, y_n)$ independent of $(x_1, \cdots, x_n)$, $\omega = \sum dx_i \wedge dy_i$ and $(x_1, \cdots, x_n)$ coordinates on the torus $\mathbb{R}^n/\mathbb{Z}^n$; such (canonical) x,y are called the action-angle variables.

But the orbits of H by itself are almost all dense 1-parameter subgroups (as soon as the $\frac{\partial H}{\partial y_k}$'s are rationally independent, for instance, in action-angle coordinates); in this case the closure of a single orbit of X_H is already an n-dimensional torus, and that's why we cannot find any more rational continuous invariants for the flow. In particular we cannot find $\pi: H \longrightarrow \mathbb{R}^{2n-1}$ which would give 2n-1 functions constant on the orbit.

A Hamiltonian vector field X_H with properties (a), (b), (c), and (d) is called a completely integrable system.

Given a completely integrable system, suppose M is the set of real points on an algebraic variety and that ω, H_i are rational differentials and functions without poles on M. Then the tori

$M_c = \pi^{-1}(c)$ are the real points on complex algebraic varieties $M_c^{\mathbb{C}}$. It may then happen, although this is a strong further assumption, that the vector fields X_{H_i} still have no poles on a compactification of $M_c^{\mathbb{C}}$. (Typically $M \subset \mathbb{R}^N$ and is the set of real points of an affine variety $M_c^{\mathbb{C}}$, leaving plenty of room for poles at infinity: e.g., take $M = \mathbb{R}^2$, $\omega = dx \wedge dy$, $H = x^4+y^4$.) If this does happen, then for a suitable complexification of the system, each M_c will be the group of real points on an <u>abelian variety</u> (or a degenerate limit which is a group formed as an extension of $(C^*)^k$ by an abelian variety). We call such systems <u>algebraically completely integrable</u>. More precisely:

<u>Definition</u>: (M^{2n},ω,H) is an algebraically completely integrable system if there exists a smooth algebraic variety $\mathfrak{M}$, a co-symplectic structure ω on $\mathfrak{M}$, i.e., $\hat{\omega} \in \Lambda^2 T_{\mathfrak{M}}$, and a morphism

$$h: \mathfrak{M} \longrightarrow \mathfrak{U}$$

$\mathfrak{U}$ a Zariski open subset of $\mathbb{C}^n$, all defined over the real field, such that

a) $\{X_i \circ h, X_j \circ h\} \underset{\text{def}}{=} \hat{\omega}(d(X_j \circ h), d(X_j \circ h)) = 0$

b) h is submersive

c) h is proper

d) M is a component of $\mathfrak{M}_{\mathbb{R}}$, the $\hat{\omega}$ on M is the $\hat{\omega}$ on $\mathfrak{M}$ along M, and H is a C^∞-function of $X_j \circ h|_M$.

In such a situation, it is easy to prove that the fibres of h are abelian varieties or extensions of these by $(\mathbb{C}^*)^k$. These remarkable cases give us methods of describing families of abelian varieties by dynamical systems.

Neumann discovered a remarkable example of this in: C. Neumann, De problemate quodam mechanico, quod ad primam integralium ultraellipticorum classem revocatur, Crelle, 56 (1859).

To describe this we start with n particles in simple harmonic oscillation, whose position is given by $x_1, \cdots, x_n$. The equations of motion are

$$\ddot{x}_i = -a_i x_i \quad ,$$

or equivalently a system of 1st order differential equations

$$\begin{cases} \dot{x}_i = y_i \\ \dot{y}_i = -a_i x_i . \end{cases}$$

We assume that $a_1 < a_2 < \cdots < a_n$, and we want to constrain the position to lie on the sphere $\sum x_i^2 = 1$; then $\sum x_i y_i = 0$, too, and the equation of motion is given by adding a force normal to S^{n-1} that keeps the particle on S^{n-1}.

We get

$$\text{(4.1)}\qquad \begin{aligned} \dot{x}_i &= y_i \\ \dot{y}_i &= -a_i x_i + x_i(\sum a_k x_k^2 - y_k^2) \end{aligned}$$

(In fact,
these imply: $\dot{(\sum x_k y_k)} = \sum y_k^2 + \sum x_k.\left(-a_k x_k + x_k(\sum a_\ell x_\ell^2 - y_\ell^2)\right)$

$$= \sum y_k^2 - \sum a_k x_k^2 + \left(\sum x_k^2\right).\left(\sum a_\ell x_\ell^2 - y_\ell^2\right)$$

which is zero if $\sum x_k^2 = 1$. Hence if we start with a point such that $\sum x_k^2 = 1$, $\sum x_k y_k = 0$, these will continue to hold if we integrate these equations.)

Let $T(S^{n-1}) = \left(\text{locus of points s.t. } \sum x_k^2 = 1,\ \sum x_k y_k = 0\right)$, i.e., the tangent bundle to S^{n-1}.

(4.1) gives a vector field on $T(S^{n-1})$:

$$D = \sum y_k \frac{\partial}{\partial x_k} - \sum a_k x_k \frac{\partial}{\partial y_k} + \left(\sum a_\ell x_\ell^2 - \sum y_\ell^2\right).\left(\sum x_k \frac{\partial}{\partial y_k}\right).$$

If we put a symplectic structure on $T(S^{n-1})$ in the usual way by restriction of the 2-form $\sum dx_i \wedge dy_i$, then (4.1) is Hamiltonian with

$$\text{(4.2)}\qquad H = \frac{1}{2}\{\sum a_k x_k^2 + \sum y_k^2\} \qquad (= \text{potential} + \text{kinetic energy}).$$

To check this, it is convenient to develop formulae which express Hamiltonian flows on symplectic submanifolds in general. Thus say

$$M \subset \mathbb{R}^{2n}$$

is defined by $f = g = 0$. Let $\omega = \sum dx_i \wedge dy_i$ define a symplectic structure on $\mathbb{R}^{2n}$ and let $\mathrm{res}_M \omega$ define one on M. We assume $\mathrm{res}_M \omega$ is non-degenerate, so ω gives us a splitting

$$T_{\mathbb{R}^{2n}}\Big|_M = T_M \oplus T_M^{\perp} .$$

$T_M^{\perp}$ is generated by the vector fields X_f, X_g, and for all functions h on $\mathbb{R}^{2n}$, the vector field $X_{\mathrm{res}(h)}$ gotten from the Hamiltonian structure on M is the projection of X_h to T_M:

$$X_{\mathrm{res}(h)} = X_h - \left(\frac{\omega(X_h, X_f)}{\omega(X_g, X_f)}\right) X_g - \left(\frac{\omega(X_h, X_g)}{\omega(X_f, X_g)}\right) X_f ,$$

where, as usual, on $\mathbb{R}^{2n}$

$$X_h = \sum \frac{\partial h}{\partial y_i} \cdot \frac{\partial}{\partial x_i} - \sum \frac{\partial h}{\partial x_i} \cdot \frac{\partial}{\partial y_i} .$$

Now, consider the special case $f = (\sum x_i^2 - 1)$, $g = \sum x_i y_i$. Then

$$X_f = -2 \sum x_i \frac{\partial}{\partial y_i}, \quad X_g = \sum x_i \frac{\partial}{\partial x_i} - \sum y_i \frac{\partial}{\partial y_i}$$

$$\omega(X_f, X_g) = 2 \sum x_i^2 = 2$$

hence

$$X_{res(h)} = \sum_k \left[\frac{\partial h}{\partial y_k} - x_k\left(\sum_i x_i \frac{\partial h}{\partial y_i}\right)\right]\frac{\partial}{\partial x_k}$$

$$+ \sum_k\left[-\frac{\partial h}{\partial x_k} + x_k\left(\sum_i x_i \frac{\partial h}{\partial x_i} - y_i \frac{\partial h}{\partial y_i}\right) + y_k\left(\sum x_i \frac{\partial h}{\partial y_i}\right)\right]\frac{\partial}{\partial y_k} .$$

Substituting $\frac{1}{2}(\sum a_k x_k^2 + \sum y_k^2)$ for h, we get (4.1.).

Following Moser, we can link these equations with Jacobians as follows: let $n = g+1$; define a map:

$$\pi: \quad T(S^g) \longrightarrow \mathbb{C}^{3g+1}$$

$$\text{by} \qquad (x,y) \longmapsto (U_{x,y}, V_{x,y}, W_{x,y})$$

where we let

$$f_1(t) = \prod_{i=1}^{n} (t-a_i)$$

$$U_{x,y}(t) = f_1(t).\sum_k \frac{x_k^2}{t-a_k} \qquad \text{monic,} \qquad \deg = n-1 = g$$

$$V_{x,y}(t) = \sqrt{-1}\ f_1(t).\sum_k \frac{x_k y_k}{t-a_k} \qquad \deg \leq n-2 = g-1$$

$$W_{x,y}(t) = f_1(t).\left(\sum_k \frac{y_k^2}{t-a_k} + 1\right) \qquad \text{monic,} \qquad \deg = n = g+1 ,$$

and the coefficients of $U_{x,y}, V_{x,y}, W_{x,y}$ are taken as the coordinates in $\mathbb{C}^{3g+1}$.

Then

$$U_{x,y}W_{x,y}+V_{x,y}^2 = f_1(t)^2\cdot\left\{\sum_{k,\ell}\frac{x_k^2y_\ell^2}{(t-a_k)(t-a_\ell)} + \sum_k \frac{x_k^2}{t-a_k} - \sum_{k,\ell}\frac{x_ky_kx_\ell y_\ell}{(t-a_k)(t-a_\ell)}\right\}$$

$$= f_1(t)^2\left\{\sum_{k<\ell}\frac{x_k^2y_\ell^2+x_\ell^2y_k^2-2x_kx_\ell y_ky_\ell}{(t-a_k)(t-a_\ell)} + \sum_k \frac{x_k^2}{t-a_k}\right\}$$

$$= f_1(t)^2\left\{\sum_{k<\ell}\frac{(x_ky_\ell-x_\ell y_k)^2}{(t-a_k)(t-a_\ell)} + \sum \frac{x_k^2}{t-a_k}\right\} \quad ;$$

because the second factor has only simple poles at a_k $\left(\text{with}\right.$ "singular part" $\left.\frac{1}{t-a_k}\left[x_k^2 + \sum_{\ell\neq k}\frac{(x_ky_\ell-x_\ell y_k)^2}{a_k-a_\ell}\right]\right)$ and is 0 at ∞ , we can re-expand by partial fractions

$$= f_1(t)^2\left\{\sum_k \frac{1}{t-a_k}\left[x_k^2 + \sum_{\ell\neq k}\frac{(x_ky_\ell-x_\ell y_k)^2}{a_k-a_\ell}\right]\right\}$$

If we set $F_k = x_k^2 + \sum_{\ell\neq k}\frac{(x_ky_\ell-x_\ell y_k)^2}{a_k-a_\ell}$

then

$$f_2(t) = \sum_k \prod_{i\neq k}(t-a_i)\cdot F_k \quad \text{is monic, of degree } n-1,$$

and finally:

$$U_{x,y}W_{x,y} + V_{x,y}^2 = f_1\cdot f_2 \quad ,$$

so that x,y defines a point of the affine point of the Jacobian of the algebraic curve $s^2 = f_1(t)\cdot f_2(t)$ embedded in $\mathbb{C}^{3g+1}$ by the method of §2!

The map π: $(x,y) \longmapsto (U,V,W)$ extends to a map $\pi_{\mathbb{C}}$ on the complexification $T(S^g)_{\mathbb{C}}$ of $T(S^g)$, i.e., the complex variety given by equations $\sum x_k^2 = 1$, $\sum x_k y_k = 0$, and the image if contained in the set of complex polynomials U,V,W such that $f_1 | V^2+UW$; or equivalently the set of affine parts of the Jacobians of the curves $s^2 = f(t)$ for which $f_1 | f$. The situation is summarized in the diagram (4.4) below.

Lemma 4.3: $\pi_{\mathbb{C}}$ is surjective; $\pi_{\mathbb{C}}(x,y) = \pi_{\mathbb{C}}(x',y')$ if and only if (x',y') is the image of (x,y) under one of the transformations $(x_k,y_k) \longmapsto (\varepsilon_k x_k, \varepsilon_k y_k)$, $\varepsilon_k = \pm 1$, which form a group of order 2^{g+1}; hence $\pi_{\mathbb{C}}$ is unramified outside the subvariety of the (U,V,W) such that $U(a_k) = V(a_k) = W(a_k) = 0$ for some k.

Proof: Given f with the property $f_1 | f$, and polynomials (U,V,W) such that $f = UW+V^2$, then we make partial fraction expansions

$$\frac{U}{f_1} = \sum_k \frac{\lambda_k}{t-a_k};$$

$$\frac{V}{f_1} = \sqrt{-1} \sum_k \frac{\mu_k}{t-a_k};$$

$$\frac{W}{f_1} = \sum_k \frac{\nu_k}{t-a_k} + 1;$$

it follows that $\sum \lambda_k = 1$ because U Monic, and it follows that $\sum \mu_k = 0$ because $\deg V \leq g-1$, and it follows that $\lambda_i \nu_i = \mu_i^2$ because at each a_i, $UW+V^2$ has a zero, hence $\frac{UW+V^2}{f_1^2}$ has a simple pole. Now we can solve for $(x_i,y_i) \in T(S^g)_{\mathbb{C}}$

$$\lambda_i = x_i^2$$
$$\nu_i = y_i^2$$
$$\mu_i = x_i y_i, \quad \text{uniquely up to a single sign for each } i.$$

QED

$$\begin{array}{ccc} & & \mathbb{C}^{3g+1} \\ & & \bigcup \ \text{Zariski-closed} \\ T(S^g)_{\mathbb{C}}^{\text{cx.dim.}2g} & \xrightarrow[\pi]{} & \left(\begin{array}{l}\text{union of the Jacobians} \\ f-V^2=U\cdot W, \text{all } f \text{ such that } f_1 \mid f\end{array}\right)^{\text{cx.dim.}2g} \\ \bigcup & & \bigcup \\ T(S^g)^{\text{real dim.}2g} & \xrightarrow[\text{res } \pi]{} & \left(\begin{array}{l}\text{subset where } f,U \text{ have} \\ \text{real roots as below, } W \\ \text{is real and } V \text{ pure imaginary}\end{array}\right)^{\text{real dim.}2g}\end{array} \tag{4.4}$$

<u>Graph of f(t):</u>

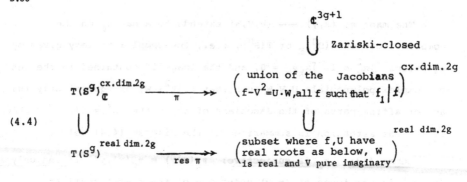

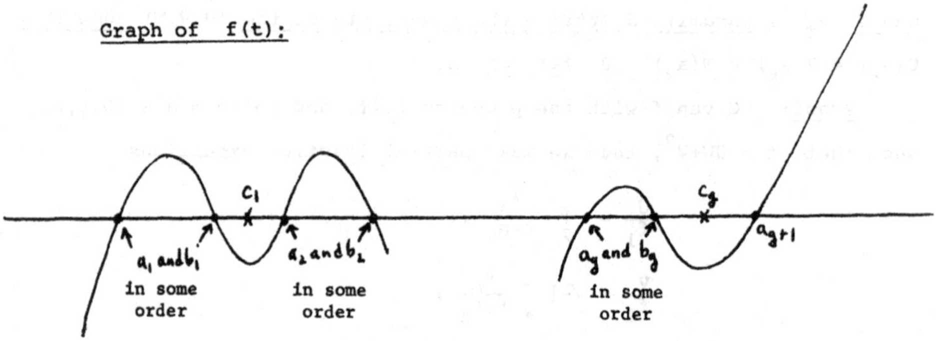

$$f_1(t) = \prod_{i=1}^{g+1}(t-a_i), \qquad f_2(t) = \prod_{i=1}^{g}(t-b_i), \qquad U(t) = \prod_{i=1}^{g}(t-c_i)$$

<u>Lemma 4.5</u>: <u>If</u> (U,V,W) <u>satisfy</u> $f_1 \mid UW+V^2$, <u>and</u> $(U,V,W) = \pi_{\mathbb{C}}(x,y)$, <u>then</u> x <u>and</u> y <u>are real if and only if</u> U,W <u>are real</u>, V <u>is pure imaginary and</u> f(t),U(t) <u>have real roots separated as in</u> (4.4).

<u>Proof</u>: If x and y are real, then

$$U(t) = \sum_k \prod_{i\neq k}(t-a_i)x_k^2$$

is a real polynomial and sign $U(a_k) = (-1)^{g-k+1}$, so U must have a zero in each of the intervals (a_k,a_{k+1}), $k \le g$. Also U is monic so $U(t) > 0$ for $t >> 0$. Thus $U(t)$ has signs like this:

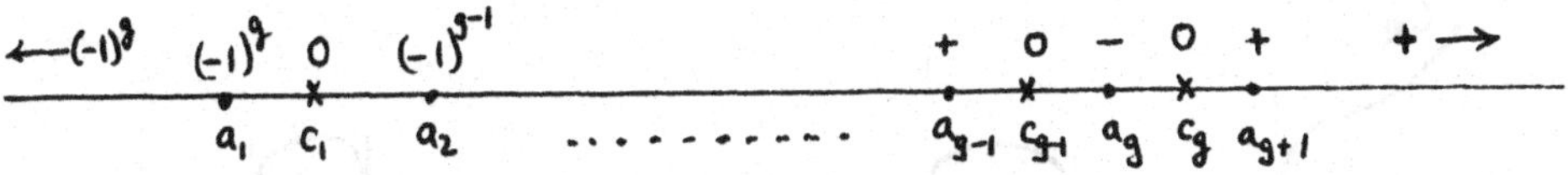

Next, $f-V^2 = U\cdot W$ and $V(a)$ is pure imaginary, all $a \in \mathbb{R}$, hence $f(t)$ is negative at all zeroes of $U(t)$, hence $f_2(t)$ is alternately + and - at these zeroes. Thus all zeroes of $f(t)$ are real with one zero of f_1 and one zero of f_2 in each interval $(-\infty,c_1),(c_1,c_2),\cdots,(c_{g-1},c_g)$ as shown in (4.4). (In all of this we have assumed the zeroes of f are distinct, but limiting cases $a_i = b_i$ and $b_i = b_{i+1}$ are possible.) Conversely, if the zeroes of U and f are real, and interweave like in the diagram, then in the partial fraction expansion above $\lambda_k \geq 0$ and $\sqrt{-1}\mu_k$ is imaginary, so the equations

$$\lambda_i = x_i^2$$

$$\nu_i = y_i^2$$

$$\mu_i = x_i y_i$$

have real solution (x,y). <u>QED</u>

If we fix f with real zeroes, the curve $s^2 = f(t)$ is a double covering of the t-line; it has a real structure given by coordinates $(s_1 = \sqrt{-1}\, s, t)$. Since $s_1^2 = -\prod_{i=1}^{g+1}(t-a_i)\cdot\prod_{i=1}^{g}(t-b_i)$, the real points on C are given by:

(4.6)

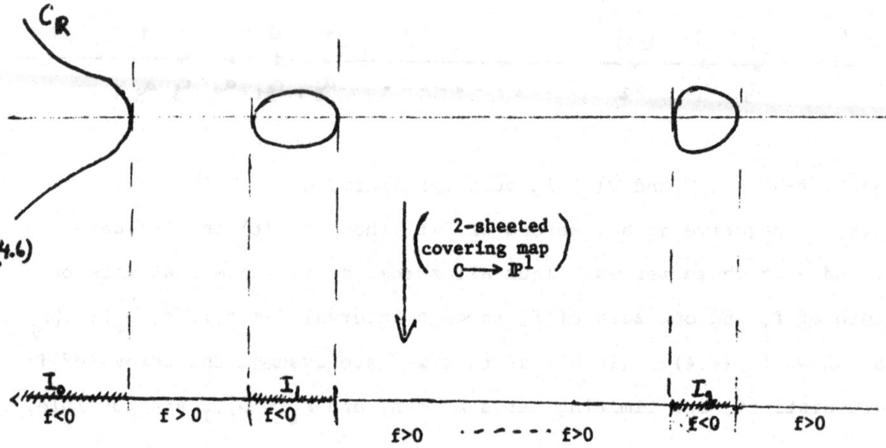

By complex conjugation $(s,t) \longmapsto (\bar{s},\bar{t})$, we get an antiholomorphic involution

$$\text{Jac } C \xrightarrow{\;-\;} \text{Jac } C$$

$(\text{Jac } C)_{\mathbb{R}}$ is defined as the set of fixed points of this involution: it is a subgroup of Jac C which must consist in a g-dimensional real subtorus plus a finite number of cosets. Since ∞ is fixed under this involution, we can determine the real points in Jac $C-\Theta$ as follows:

$$(\text{Jac } C)_{\mathbb{R}} - \Theta = \left\{ \sum_{i=1}^{g} P_i - g\infty \;\middle|\; P_i \neq \infty, P_i \neq \iota P_j, \text{ if } i \neq j, \text{ and } \sum \bar{P}_i = \sum P_i \right\}.$$

Note that $\sum \bar{P}_i = \sum P_i$ means that $\sum P_i$ consists in some real points and some pairs of conjugate complex points. If $(I_0,\cdots,I_g)$ are the intervals in $\mathbb{R}$ where $f \leq 0$ (see diagram (4.6)), then any subset S of $\{0,1,\cdots,g\}$ whose cardinality is $\leq g$ and $\equiv g \bmod 2$ defines a connected component K_S of $(\text{Jac } C)_{\mathbb{R}} - \Theta$, namely the set of the divisor classes $\sum_{i=1}^{g} P_i - g\infty$ such that, if $S = \{i_1,\cdots,i_s\}$,

$$K_S \begin{cases} a)\ t(P_k) \in I_{i_k}, & \text{so } \overline{P}_k = P_k \text{ , for } 1 \le k \le s \\ \text{and} & \\ b)\ \overline{P}_{s+1} = P_{s+2} & \underline{\text{or}}\ \ t(P_{s+1}), t(P_{s+2}) \in (\text{same } I_{\ell_1}) \\ \text{and} & \\ c)\ \overline{P}_{s+3} = P_{s+4} & \underline{\text{or}}\ \ t(P_{s+3}), t(P_{s+4}) \in (\text{same } I_{\ell_2}). \\ \quad \text{etc.} & \end{cases}$$

<u>Example:</u>

$g = 2$,

the real (affine) part of the Jacobian breaks up into 4 components; in order for $\sum_{i=1}^{2} P_i$ to be equal to $\sum_{i=1}^{2} \overline{P}_i$, the only possibilities are:

$K_{\{0,1\}}$: P_1 over I_0, P_2 over I_1

$K_{\{0,2\}}$: P_1 over I_0, P_2 over I_2

$K_{\{1,2\}}$: P_1 over I_1, P_2 over I_2

$K_\emptyset$: $(P_1, P_2$ over the same $I_K)$ or $(P_2 = \overline{P}_1)$.

Note that this last set $K_\emptyset$ is connected because we can continuously move $P+\overline{P}$ on the complex curve C until $P = \overline{P} \in I_k$ and then move the two points independently on the real loop over I_k; $K_\emptyset$ also contains 2∞ in the limit, hence it's the connected component of the origin. Thus we verify that when $g = 2$,

$(\text{Jac } C)_{\mathbb{R}}$ is a real 2-dimensional closed subgroup,

isomorphic to $\mathbb{R}^2/\text{lattice} \times (\mathbb{Z}/2\mathbb{Z})^2$.

Returning to Neumann's dynamical system, the following theorem will show that the functions $F_k(x,y)$ are integrals of the vector field X_H give commuting flows on $T(S^g)_{\mathbb{C}}$ and show that their image under π is tangent to all Jacobians, and gives the translation-invariant flows on them.

Theorem 4.7 (Moser-Uhlenbeck*). On $T(S^g)_{\mathbb{C}}$

a) $\{F_k, F_\ell\} = 0$

b) $\sum_{k=1}^{g+1} F_k = \sum x_k^2 = 1, \quad \frac{1}{2}\sum_{k=1}^{g+1} a_k F_k = H$

c) $\pi_*(X_{F_k}) = c_k D_{a_k}, \quad c_k = 4\sqrt{-1} \cdot \prod_{\ell \neq k} (a_k - a_\ell)^{-1}$

$\pi_*(X_H) = cD_\infty, \qquad c = -2\sqrt{-1}$

d) the X_{F_k} span a g-dimensional space, except over the Zariski closed subset of triples (U,V,W) which have a common root.

Proof of a).

First one checks that on $\mathbb{C}^{2g+2}$, with coordinates x_i, y_i, $\{F_k, F_\ell\} = 0$ with respect to the symplectic form $\sum dx_i \wedge dy_i = \omega$; if we let $z_{k\ell} = x_k y_\ell - y_k x_\ell$, then

$$\{z_{k\ell}, z_{ij}\} = \sum_\alpha -\frac{\partial z_{k\ell}}{\partial x_\alpha}\frac{\partial z_{ij}}{\partial y_\alpha} + \frac{\partial z_{k\ell}}{\partial y_\alpha}\frac{\partial z_{ij}}{\partial x_\alpha}, \quad \text{hence}$$

$$\{z_{k\ell}, z_{ij}\} = 0 \qquad \text{if} \quad \{k,\ell\} \cap \{i,j\} = \phi$$

$$\{z_{k\ell}, z_{kj}\} = -y_\ell(-x_j) + (-x_\ell)y_j = -z_{\ell j}$$

---

*J. Moser, Various aspects of hamiltonian systems, C.I.M.E. conference talks, Bressone, 1978; K. Uhlenbeck, Equivariant harmonic maps into spheres, Proc. Tulane Conf. on Harmonic Maps.

Thus if $\phi_k = \sum_{j\neq k} \frac{z_{jk}^2}{a_k - a_j}$ and $k \neq \ell$

$$\{\phi_k,\phi_\ell\} = \sum_{\substack{j\neq k\\ i\neq \ell}} \frac{1}{a_k-a_j}\,\frac{1}{a_\ell-a_i}\,\{z_{kj}^2, z_{\ell i}^2\} = 4 \sum_{\substack{j\neq k\\ i\neq \ell}} \frac{1}{a_k-a_j}\,\frac{1}{a_\ell-a_i}\, z_{kj} z_{\ell i}\{z_{kj}, z_{\ell i}\}$$

$$= 4 \sum_{\substack{j\neq k\\ i\neq \ell}} \frac{1}{a_k-a_j}\,\frac{1}{a_\ell-a_i} \cdot \begin{cases} -z_{k\ell} z_{kj}\, z_{\ell j} & \text{if } i = j \\ z_{k\ell} z_{\ell i}\, z_{ki} & \text{if } j = \ell \\ z_{kj} z_{\ell k}\, z_{j\ell} & \text{if } i = k \end{cases}$$

$$= -4\, z_{k\ell} \sum_{\substack{i\neq k\\ \text{or } \ell}} \frac{1}{a_k-a_i}\,\frac{1}{a_\ell-a_i}\, z_{ki}\, z_{\ell i} \; +$$

$$+\; 4\, z_{k\ell}\, \frac{1}{a_k-a_\ell} \cdot \sum_{i\neq \ell} \frac{z_{ki}\, z_{\ell i}}{a_\ell-a_i} + 4 z_{k\,\ell}\, \frac{1}{a_\ell-a_k} \cdot \sum_{j\neq k} \frac{z_{\ell j} z_{kj}}{a_k-a_j}$$

$$= -4\, z_{k\ell} \sum_{\substack{i\neq k\\ \text{or } \ell}} \left[\frac{1}{(a_k-a_i)(a_\ell-a_i)} - \frac{1}{(a_k-a_\ell)(a_\ell-a_i)} - \frac{1}{(a_\ell-a_k)(a_k-a_i)}\right] z_{ki} z_{\ell i}$$

$$= -4\, z_{k\ell} \sum \frac{(a_k-a_\ell)-(a_k-a_i)+(a_\ell-a_i)}{(a_k-a_i)(a_\ell-a_i)(a_k-a_\ell)}\, z_{ki} z_{\ell i}$$

$$= 0 \quad .$$

Finally

$$\{F_k,F_\ell\} = \{x_k^2+\phi_k,\; x_\ell^2+\phi_\ell\}$$

$$= \{x_k^2, x_\ell^2\} + \{x_k^2, \phi_\ell\} + \{\phi_k, x_\ell^2\}$$

$$= -2x_k\, \frac{\partial \phi_\ell}{\partial y_k} + 2x_\ell\, \frac{\partial \phi_k}{\partial y_\ell}$$

$$= 2x_k \frac{2(x_k y_\ell - x_\ell y_k)(-x_\ell)}{a_k - a_\ell} - 2x_\ell \frac{2(x_\ell y_k - x_k y_\ell)(-x_k)}{a_\ell - a_k}$$

$$= 0 .$$

To conclude use the following

<u>Lemma 4.8</u>: <u>Let</u> $M^{2n} \supset N^{2n-2}$ <u>be symplectic with basis</u> 2-<u>forms</u> ω, res ω; <u>let</u> f,g,h <u>be functions on</u> M <u>such that</u> df <u>is nowhere zero and</u> f = 0 <u>on</u> N. <u>Then if the Poisson brackets on</u> M <u>satisfy</u> $\{f,g\}_M = \{g,h\}_M = 0$, <u>the Poisson bracket on</u> N <u>satisfies</u> $\{g,h\}_N = 0$.

<u>Proof</u>: Via $\hat{\omega}$, we can split the cotangent bundle to M into the cotangent bundle to N and its orthogonal complement

$$T_M^* = T_N^* \oplus C^* .$$

Now write

$$dg = d(\text{res } g) + \alpha$$
$$dh = d(\text{res } h) + \beta$$
$$df = \gamma \neq 0 \qquad \alpha,\beta,\gamma \in C^* .$$

From $\hat{\omega}(\alpha,\gamma) = \hat{\omega}(\beta,\gamma) = 0$ it follows that $\alpha = a\cdot\gamma$, $\beta = b\cdot\gamma$ are linearly dependent, because C^* is 2-dimensional, so $\hat{\omega}(\alpha,\beta) = 0$, so $0 = \hat{\omega}(dg,dh) = \hat{\omega}(d \text{ res } g, d \text{ res } h) = \{\text{res } g, \text{res } h\}$. <u>QED</u>

<u>Proof of b)</u>.

$$\sum F_k = \sum x_k^2 + \sum_{\substack{k,\ell \\ k\neq\ell}} \frac{(x_k y_\ell - x_\ell y_k)^2}{a_k - a_\ell}$$

$$= \sum x_k^2 + \sum_{k<\ell} (x_k y_\ell - x_\ell y_k)^2 \left(\frac{1}{a_k - a_\ell} + \frac{1}{a_\ell - a_k}\right)$$

$$= 1$$

and

$$\frac{1}{2}\sum a_k F_k = \frac{1}{2}\sum a_k x_k^2 + \frac{1}{2}\sum_{\substack{k,\ell\\k\neq\ell}} a_k \frac{(x_k y_\ell - x_\ell y_k)^2}{a_k - a_\ell}$$

$$= \frac{1}{2}\sum a_k x_k^2 + \frac{1}{2}\sum_{k<\ell}\left(\frac{a_k}{a_k-a_\ell} + \frac{a_\ell}{a_\ell - a_k}\right)(x_k y_\ell - x_\ell y_k)^2$$

$$= \frac{1}{2}\sum a_k x_k^2 + \frac{1}{4}\sum_k\sum_\ell (x_k y_\ell - x_\ell y_k)^2$$

$$= \frac{1}{2}\sum a_k x_k^2 + \frac{1}{4}\left[2\sum_k x_k^2\cdot\sum_k y_k^2 - 2\left(\sum_k x_k y_k\right)^2\right]$$

$$= \frac{1}{2}\sum a_k x_k^2 + \frac{1}{2}\sum y_k^2 = H.$$

Proof of c): Under the F_k-flow on $T(S^g)_{\mathbb{C}}$, by the formulae for Hamiltonian flows on a submanifold:

$$\dot{x}_\ell = \frac{\partial F_k}{\partial y_\ell} - x_\ell\left(\sum_p x_p \frac{\partial F_k}{\partial y_p}\right).$$

Since

$$\sum_p x_p \frac{\partial F_k}{\partial y_p} = \sum_{p\neq k} x_p \frac{2(x_k y_p - x_p y_k)x_k}{a_k - a_p} + x_k \sum_{p\neq k}\frac{2(x_k y_p - x_p y_k)(-x_p)}{a_k - a_p} = 0$$

then

$$\dot{x}_\ell = \frac{2(x_k y_\ell - x_\ell y_k)}{a_k - a_\ell}\, x_k \qquad \text{if } \ell \neq k$$

or

$$= \sum_{p\neq k}\frac{2(x_k y_p - x_p y_k)}{a_k - a_p}(-x_p) \qquad \text{if } \ell = k.$$

Under the action of π, this vector field becomes

$$\dot{\lambda}_\ell = (\dot{x_\ell^2}) = 2x_\ell\dot{x}_\ell = 4\,\frac{x_k x_\ell(x_k y_\ell - x_\ell y_k)}{a_k - a_\ell}$$

$$= 4\,\frac{\lambda_k\mu_\ell - \lambda_\ell\mu_k}{a_k - a_\ell} \qquad \text{if } \ell \neq k$$

$$\text{or} \quad = \sum_{p\neq k} \frac{-4x_p x_k(x_k y_p - x_p y_k)}{a_k - a_p} \qquad \text{if } \ell = k$$

$$= -4\sum_{p\neq k} \frac{\lambda_k\mu_p - \lambda_p\mu_k}{a_k - a_p} \quad .$$

So $\dot{U}(t) = f_1(t)\sum \frac{\dot{\lambda}_\ell}{t-a_\ell}$

$$= f_1(t)\left[4\sum_{\ell\neq k}\frac{\lambda_k\mu_\ell - \lambda_\ell\mu_k}{a_k - a_\ell}\,\frac{1}{t-a_\ell} - 4\sum_{p\neq k}\frac{\lambda_k\mu_p - \lambda_p\mu_k}{a_k - a_p}\,\frac{1}{t-a_k}\right]$$

$$= f_1(t)\left[-4\sum_{\ell\neq k}\frac{\lambda_k\mu_\ell - \lambda_\ell\mu_k}{(t-a_\ell)(t-a_k)}\right]$$

$$= -4\left[f_1(t)\,\frac{\lambda_k}{t-a_k}\sum_\ell\frac{\mu_\ell}{t-a_\ell} - \frac{\mu_k}{t-a_k}\sum_\ell\frac{\lambda_\ell}{t-a_\ell}\right]$$

$$= -4\left[\frac{\lambda_k V(t)/\sqrt{-1} - \mu_k U(t)}{t - a_k}\right].$$

But

$$U(t) = \sum_k \prod_{\ell\neq k}(t-a_\ell)\lambda_k$$

$$V(t) = \sqrt{-1}\sum_k\prod_{\ell\neq k}(t-a_\ell)\mu_k$$

so $$U(a_k) = \prod_{\ell\neq k}(a_k-a_\ell)\cdot\lambda_k,\quad V(a_k) = \sqrt{-1}\prod_{\ell\neq k}(a_k-a_\ell)\mu_k$$

and finally $$\dot{U}(t) = \sqrt{-1}\cdot 4\left[\frac{V(a_k)U(t)-U(a_k)V(t)}{t-a_k}\right]\prod_{\ell\neq k}(a_k-a_\ell)^{-1}$$

$$=\left(4\sqrt{-1}\cdot\prod_{\ell\neq k}(a_k-a_\ell)^{-1}\right)\cdot D_{a_k}(U(t)).$$

The argument at the end of the proof of theorem 3.1 shows that for any vector X on the space of (U,V,W)'s which is tangent to the fibre $UW+V^2 = f$ at a point where U,V have no common zeroes, the equality

$$X(U(t)) = c\cdot\frac{U(t)V(P)-U(P)V(t)}{t-t(P)},\quad \text{c a constant, implies}$$

$$X(V(t)) = \frac{c}{2}\cdot\left[\frac{U(P)W(t)-W(P)U(t)}{t-t(P)} - U(P)U(t)\right]$$

$$X(W(t)) = c\cdot\left[\frac{W(P)V(t)-V(P)W(t)}{t-t(P)} - U(P)V(t)\right].$$

So in order to finish the proof of c)*, where the constant is $c = 4\sqrt{-1}\cdot\prod_{\ell\neq k}(a_k-a_\ell)^{-1}$, we notice that by a) the collection of vector fields X_{F_k} are tangent to the loci (F_ℓ = constant, all ℓ), hence their images via π_* are tangent to the fibres sitting over $f = f_1\cdot\sum_k F_k\cdot\prod_{\ell\neq k}(t-a_\ell)$. Thus we get $\pi_* X_{F_k} = cD_{a_k}$ on the part of the fibre where U,V are relatively prime; the result holds everywhere by continuity.

The formula for $\pi_*(X_H)$ is proven similarly, the calculation being much simpler.

---

*Alternatively, one may calculate $\dot{V}(t),\dot{W}(t)$ directly as we did $\dot{U}(t)$.

<u>Proof of d)</u>. Assume that at a particular triple (U,V,W), there are constants d_i such that $\sum d_i D_{a_i} = 0$. Then

$$\sum_{i=1}^{g} d_i \cdot \frac{V(a_i)U(t)-U(a_i)V(t)}{t - a_i} \equiv 0$$

or

$$\left(\sum_{i=1}^{g} d_i \cdot \frac{V(a_i)}{t-a_i}\right)U(t) = \left(\sum_{i=1}^{g} d_i \cdot \frac{U(a_i)}{t-a_i}\right) V(t)$$

or

$$\underbrace{\left[\sum d_i\ V(a_i) \prod_{j\neq i} (t-a_j)\right]}_{V^*}U(t) = \underbrace{\left[\sum d_i\ U(a_i) \prod_{j\neq i} (t-a_j)\right]}_{U^*}V(t) \quad ;$$

since deg U = g, and deg $U^* \leq g-1$, this implies that U and V have a common root a where $U^*(a) \neq 0$ (or at least the multiplicity of a as a root of U^* is less than its multiplicity for U).

Likewise

$$\sum_{i=1}^{g} d_i \left[\frac{U(a_i)W(t)-W(a_i)U(t)}{t - a_i} - U(a_i)U(t)\right] = 0$$

or

$$\left(\sum d_i \frac{U(a_i)}{t-a_i}\right) W(t) = \left[\sum d_i \frac{W(a_i)}{t-a_i} + \sum d_i\ U(a_i)\right]U(t)$$

implies

$$\underbrace{\Big[\sum d_i\ U(a_i)\prod_{j\neq i}(t-a_j)\Big]}_{U^*}W(t) = \underbrace{\Big[\sum d_i W(a_i)\prod_{j\neq i}(t-a_j)+\sum d_i U(a_i)\prod_j(t-a_j)\Big]}_{W^*}U(t)$$

hence $U(a) = 0$, $U^*(a) \neq 0$ implies $W(a) = 0$ too. Thus U,V,W have a common root.

Corollary 4.9: Almost all orbits of $\{X_{F_k}\}$ (defined by F_k = const., all k) are compact real tori, isomorphic to connected components of the real points on a 2^{g+1}-order covering of the Jacobian of a hyperelliptic curve.

The covering that occurs here will be described analytically in §5.

Finally, Moser discovered a beautiful link between the dynamical system $T(S^{n-1}),\{F_k\}$ and the problem of finding the geodesics on an ellipsoid. The result is so elegant that we want to reproduce it here:

Theorem 4.10 (Moser). Let E be the ellipsoid $\sum \frac{x_k^2}{a_k} = 1$ in $\mathbb{R}^{g+1}$. If $\vec{x},\vec{y} \in \mathbb{R}^{g+1}$ satisfy $|\vec{x}| = 1$, $\langle\vec{x},\vec{y}\rangle = 0$ then:

$$\left(\sum \frac{1}{a_k} F_k(x,y) = 0\right) \text{ if and only if } \left(\begin{array}{c}\text{the line } L_{x,y}=\{\vec{y}+t\vec{x}\mid t\in\mathbb{R}\}\\ \text{is tangent to } E\end{array}\right),$$

and if this holds:

$$\vec{\xi} = \left[\vec{y} - \frac{\left(\sum \frac{x_k y_k}{a_k}\right)}{\left(\sum \frac{x_k^2}{a_k}\right)}\cdot\vec{x}\right] = L_{x,y}\cap E.$$

If $\vec{x}(t),\vec{y}(t)$ is an integral curve for the vector field $\sum \frac{1}{a_k} F_k$, then $\vec{\xi}(t)$ is a geodesic on E, up to reparametrization.

Proof: First we calculate out

$$\sum \frac{1}{a_k} F_k(x,y) = \sum \frac{1}{a_k}\left(x_k^2 + \sum_{\ell \neq k} \frac{(x_k y_\ell - x_\ell y_k)^2}{a_k - a_\ell}\right)$$

$$= \sum \frac{x_k^2}{a_k} + \sum_{\ell > k} \left(\frac{1}{a_k} - \frac{1}{a_\ell}\right) \frac{(x_k y_\ell - x_\ell y_k)^2}{a_k - a_\ell}$$

$$= \sum \frac{x_k^2}{a_k} - \frac{1}{2} \sum_{\ell,k} \frac{(x_k y_\ell - x_\ell y_k)^2}{a_k a_\ell}$$

$$= \sum \frac{x_k^2}{a_k} - \sum_k \frac{x_k^2}{a_k} \cdot \sum_\ell \frac{y_\ell^2}{a_\ell} + \left(\sum \frac{x_k y_k}{a_k}\right)^2$$

$$= \sum \frac{x_k^2}{a_k}\left(1 - \sum \frac{y_k^2}{a_k}\right) + \left(\sum \frac{x_k y_k}{a_k}\right)^2 ,$$

or if $B(u,v)$ is the bilinear form $\sum u_k v_k / a_k$,

$$\sum \frac{1}{a_k} F_k(x,y) = B(x,x)(1 - B(y,y)) + B(x,y)^2.$$

Call this function F. We calculate the flow associated to F.

a) $$\sum x_k \frac{\partial F}{\partial y_k} = B(x,x) \sum_k (-2x_k)\frac{y_k}{a_k} + 2B(x,y) \sum_k x_k (x_k/a_k)$$

$$= -2B(x,x)B(x,y) + 2B(x,y)B(x,x) = 0 .$$

b) $\sum_k x_k \frac{\partial F}{\partial x_k} = 2F$ because F is homogeneous in x of degree 2.

c) Likewise $\sum y_k \frac{\partial F}{\partial y_k}$ picks out the quadratic y-terms in F, i.e.,

$$\sum y_k \frac{\partial F}{\partial y_k} = -2B(x,x)B(y,y) + 2B(x,y)^2,$$

so

$$\sum x_k \frac{\partial F}{\partial x_k} - \sum y_k \frac{\partial F}{\partial y_k} = 2B(x,x).$$

d) The flow therefore is

$$\dot{x}_k = \frac{\partial F}{\partial y_k} = -2\frac{y_k}{a_k}B(x,x) + 2\frac{x_k}{a_k}B(x,y)$$

$$\dot{y}_k = -\frac{\partial F}{\partial x_k} + x_k(2B(x,x))$$

$$= -2\frac{x_k}{a_k}(1 - B(y,y)) - 2\frac{y_k}{a_k}B(x,y) + 2x_k B(x,x).$$

Let E be the ellipsoid $B(x,x) = 1$. Note that $B(\vec{y}+t\vec{x},\vec{y}+t\vec{x})$ has a minimum at $t = -B(x,y)/B(x,x)$, i.e., at $B(\vec{\xi},\vec{\xi})$ if $\vec{\xi} = \vec{y} - \frac{B(x,y)}{B(x,x)}\vec{x}$. But

$$B(\xi,\xi) = B(y,y) - \frac{B(x,y)^2}{B(x,x)}.$$

So $L_{x,y}$ is tangent to E if and only if $B(\xi,\xi) = 1$, which holds if and only if $1-B(y,y) + B(x,y)^2/B(x,x) = 0$, i.e., if and only if $F(x,y) = 0$. Now differentiating along a flow line:

$$\dot{\xi}_k = \dot{y}_k - \frac{B(x,y)}{B(x,x)}\dot{x}_k - \left(\frac{B(x,y)}{B(x,x)}\right)^{\cdot} \cdot x_k$$

$$= -2\frac{x_k}{a_k}(1-B(y,y)) - 2\frac{y_k}{a_k}B(x,y) + 2x_k B(x,x)$$

$$+ 2\frac{y_k}{a_k}B(x,y) - 2\frac{x_k}{a_k}\frac{B(x,y)^2}{B(x,x)} - \left(\frac{B(x,y)}{B(x,x)}\right)^{\cdot} \cdot x_k .$$

Now define a function $\tau(t)$ by setting

$$\tau(t)^{\cdot} = 2B(x,x) - \left(\frac{B(x,y)}{B(x,x)}\right)^{\cdot}$$

along this flow line. Then

$$\frac{d\xi_k}{dt} = \frac{d\tau}{dt}\cdot x_k$$

or

$$\frac{d\xi_k}{d\tau} = x_k .$$

Therefore

$$\frac{d^2\xi_k}{d\tau^2} = \frac{d}{d\tau}\, x_k$$

$$= \frac{1}{d\tau/dt}\cdot \dot{x}_k$$

$$= \frac{1}{d\tau/dt}\left[\frac{y_k - \frac{B(x,y)}{B(x,x)}\, x_k}{a_k}\right](-2\, B(x,x))$$

$$= -\frac{2B(x,x)}{d\tau/dt}\cdot\frac{\xi_k}{a_k} .$$

This simply says that the acceleration of $\vec{\xi}(\tau) \in E$ is always normal to the ellipsoid E, i.e., that $\vec{\xi}(\tau)$ is a geodesic. QED

§5. Tying together the analytic Jacobian and algebraic Jacobian

So far in this Chapter, we have defined an algebraic variety Jac C and studied its invariant flows. In Chapter II, we associated to any compact Riemann Surface C a complex torus Jac C. If C is hyperelliptic so that both constructions apply, they are isomorphic by Abel's theorem. We would now like to make this isomorphism explicitly, i.e., express the algebraic coordinates on Jac C-Θ as theta functions.

To study C as we did in Chapter II, the first thing we must do is to choose a homology basis A_i, B_i. There is a traditional way to do this in the hyperelliptic case. One first chooses on $\mathbb{P}^1 = \mathbb{C} \cup (\infty)$, a simple closed curve ρ through the set of branch points B. One then chooses paths in $\mathbb{P}^1$-B as in the diagram below. Noting that each of them circles an even number of branch points, these paths can be lifted to the double cover C: see Figure on next page.

On C, the paths A_i are disjoint from each other as are the paths B_i, and A_i, B_j meet only if $i = j$, and then in one point so that

$$i(A_i, A_j) = i(B_i, B_j) = 0$$

$$i(A_i, B_j) = \delta_{ij}.$$

Thus A_i, B_j are a symplectic basis of $H_1(C, \mathbb{Z})$. To make this picture clearly homeomorphic to the figure in Ch. II, §2, we can also add disjoint tails to all A_i, B_i, connecting them to the base point P_1. Widening each tail into 4 parallel paths, we can lengthen A_i, B_i to disjoint simple closed loops A_i', B_i' all beginning and ending at P_1, which is exactly as in §II.2.

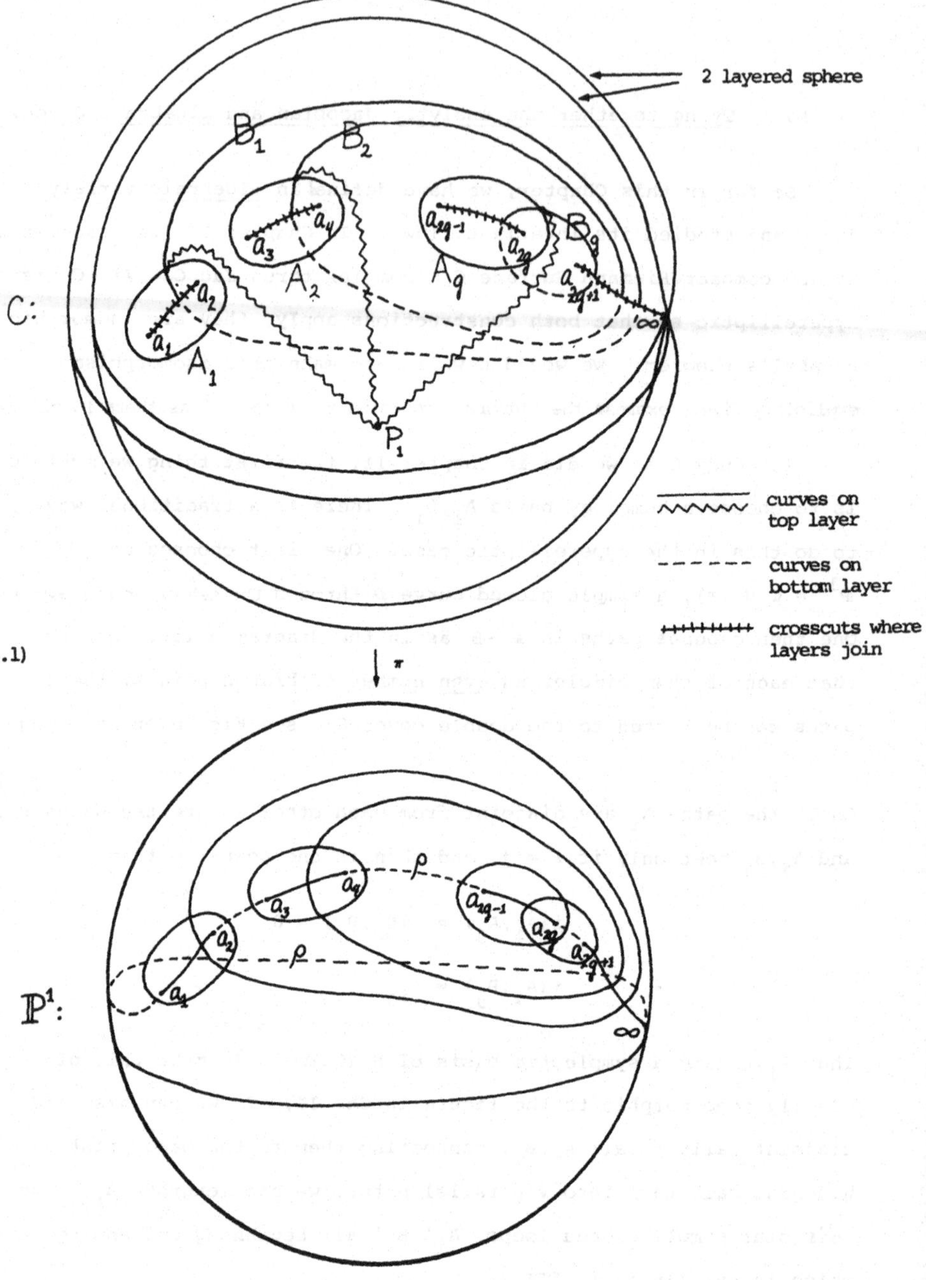
2 layered sphere
C:
B1
B2
Bg
A1
A2
Ag
a1
a2
a3
a4
a2g-1
a2g
a2g+1
∞
P1
curves on top layer
curves on bottom layer
crosscuts where layers join
π
P1:
ρ
ρ
a1
a2
a3
a4
a2g-1
a2g
a2g+1
∞

(5.1)

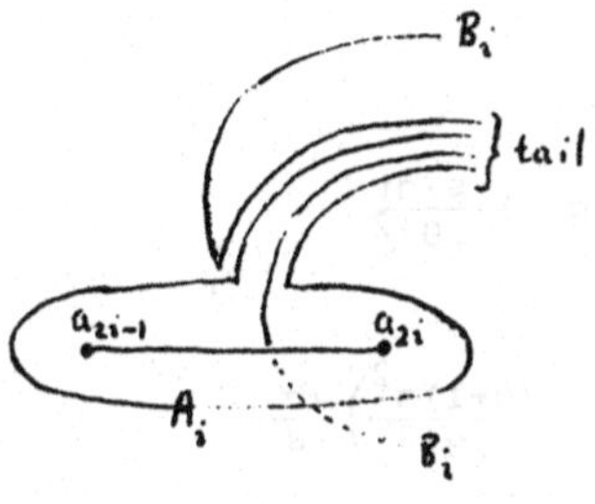

Next, on C we can describe the g-dimensional vector space of holomorphic 1-forms:

Proposition 5.2: $\Gamma(C,\Omega^1)$ consists in the 1-forms:

$$\omega = \frac{P(t)dt}{s}, \quad P \text{ a polynomial of degree } \leq g-1.$$

Proof: Because $s^2 = f(t)$, we have

$$2s\,ds = f'(t)\cdot dt$$

so

$$\frac{P(t)dt}{s} = \frac{2P(t)ds}{f'(t)}.$$

On C_1 (the affine piece of C with coordinates s,t), $s = 0$ implies $f(t) = 0$ which implies $f'(t) \neq 0$ because f has no double roots. Thus at every $P \in C_1$, either $s(P) \neq 0$ or $f'(t(P)) \neq 0$, so using one of the above expressions it follows that ω has no poles on C_1. Now at ∞, $t' = \frac{1}{t}$ and $s' = \frac{s}{t^{g+1}}$ are coordinates. Then

$$ds' = \frac{ds}{t^{g+1}} - (g+1)\,\frac{s\cdot dt}{t^{g+2}}$$

$$= \left(\frac{f'(t)}{2\cdot t^{g+1}} - \frac{(g+1)s^2}{t^{g+2}}\right)\frac{dt}{s}$$

$$= \frac{tf'(t)-(2g+2)f(t)}{2t^{g+2}}\cdot\frac{dt}{s}$$

$$= \frac{-t^{2g+1}+(\text{lower terms in } t)}{2t^{g+2}}\cdot\frac{dt}{s}.$$

Now we saw in §1 that s',t' have respectively a simple and a double zero at ∞ ; hence s' is a local coordinate near ∞ and ds' is a 1-form with neither zero nor pole at ∞. So the above equation shows that

$$\frac{dt}{s} = (-2(t')^{g-1} + \text{higher order terms in } t')\cdot ds',$$

i.e., $\frac{dt}{s}$ has a zero of order $2g-2$ at ∞. Thus if $\deg P \leq g-1$, $P(t)$ has pole at ∞ of order $\leq 2g-2$ and ω has no poles at all. Thus we have found a g-dimensional space of 1-forms without poles and as $\dim \Gamma(C,\Omega^1) = g$, this must be all of them. (We could also start with an arbitrary rational 1-form $\eta = (\phi(t)+s\psi(t))dt$ and show directly that if η has no poles, then $\phi = 0$, $\psi(t) = (\text{polyn. of deg} \leq g-1)/f(t)$.) QED

The next step is to choose a normalized basis

$$\omega_i = \frac{P_i(t)dt}{s}$$

of holomorphic 1-forms such that

$$\int_{A_i} \omega_j = \delta_{ij} .$$

The period matrix of the curve C is then

$$\Omega_{ij} = \int_{B_i} \omega_j ,$$

and the analytic Jacobian is by definition:

$$\mathbb{C}^g/L_\Omega, \qquad L_\Omega = (\text{lattice } \mathbb{Z}^g + \Omega\cdot\mathbb{Z}^g).$$

By means of the indefinite abelian integrals we have holomorphic maps

$$\begin{array}{ccc} C^k & \xrightarrow{\quad I \quad} & \mathbb{C}^g/L_\Omega \\ (P_1,\cdots,P_k) & \longmapsto & \left(\sum \int_\infty^{P_i} \vec{\omega}\right) \bmod L_\Omega . \end{array}$$

Abel's theorem (II, §2) states that these induce an isomorphism

$$\text{Jac } C \xrightarrow{\approx} \mathbb{C}^g/L_\Omega$$

if we map a divisor class $\sum_1^k P_i - \sum_1^k Q_i$ to $I(P_1,\cdots,P_k) - I(Q_1,\cdots,Q_k)$. Taking $k = g$, we compare this with our algebraic description of an affine piece of Jac C:

$$
\begin{array}{ccc}
\mathbb{C}^g & \xrightarrow{\quad I \quad} \!\!\!\!\to & \mathbb{C}^g/L_\Omega \\
\cup & & \\
(\mathbb{C}^g)_o & \xrightarrow{\quad \sim \quad} \!\!\!\!\to & Z \subset \mathbb{C}^{2g} \\
\| & & \| \\
\left(\begin{array}{l}\text{open set of } P_1,\cdots,P_g, \\ P_i \neq \infty, P_i \neq \iota P_j \text{ if } i \neq j\end{array}\right) & & \left(\begin{array}{l}\text{variety of polyn. } U,V \\ \text{such that } U \mid f-V^2\end{array}\right)
\end{array}
$$

We have seen that Z is the open piece Jac $C-\Theta$ of the Jacobian, where Θ = (locus of divisor classes $\sum_{i=1}^{g-1} P_i - (g-1)\cdot\infty$). Our goal now is to prove:

<u>Theorem 5.3</u>:

1) <u>There are</u> $\vec{\delta}', \vec{\delta}'' \in \frac{1}{2}\mathbb{Z}^g$ <u>such that for all</u> $\vec{z} \in \mathbb{C}^g$,

$$
\vartheta\begin{bmatrix}\vec{\delta}' \\ \vec{\delta}''\end{bmatrix}(\vec{z},\Omega) = 0 \iff \left(\begin{array}{c}\exists P_1,\cdots,P_{g-1} \in C \ \underline{\text{such that}} \\ \vec{z} \equiv \sum_{i=1}^{g-1} \int_\infty^{P_i} \vec{\omega} \mod L_\Omega\end{array}\right)
$$

2) <u>Thus</u> Jac $C-\Theta$ <u>can be described analytically as</u> $\left(\mathbb{C}^g/L_\Omega\right) - V\left(\vartheta\begin{bmatrix}\delta' \\ \delta''\end{bmatrix}\right)$, <u>and algebraically as the above variety</u> Z, <u>whose coordinates are the coefficients of</u> $U(t), V(t)$. <u>Thus the coefficients of</u> $U(t), V(t)$ <u>are meromorphic functions on</u> $\mathbb{C}^g/L_\Omega$ <u>with poles where</u> $\vartheta\begin{bmatrix}\delta' \\ \delta''\end{bmatrix} = 0$.

3) For all branch points $a_k \in B$, there are $\vec{\eta}'(k), \vec{\eta}''(k) \in \frac{1}{2}\mathbb{Z}^g$ and a constant c_k such that for all divisors $D = \sum_{i=1}^{g} P_i$ in $(C^g)_o$, if $U^D(t) = \prod_{i=1}^{g}(t-t(P_i))$ is the corresponding polynomial, then

$$U^D(a_k) = c_k \cdot \left[\frac{\vartheta\begin{bmatrix}\delta'+\eta_k'\\ \delta''+\eta_k''\end{bmatrix}\left(\sum_{i=1}^{g}\int_{\infty}^{P_i}\vec{\omega}\right)}{\vartheta\begin{bmatrix}\delta'\\ \delta''\end{bmatrix}\left(\sum_{i=1}^{g}\int_{\infty}^{P_i}\vec{\omega}\right)} \right]^2$$

This determines the coefficients of $U(t)$ as meromorphic functions on $\mathbb{C}^g/L_\Omega$.

In the course of proving this, we shall determine $\delta, \eta(k)$ explicitly. In fact, c_k can also be determined, but this will not be done until the next section.

We first prove (1). Note that (1) is exactly Corollary 3.6 of Riemann's Theorem, (Ch. II, §3), except that we assert that $\vec{\Delta} \in \frac{1}{2}L_\Omega$ and we want to compute $\vec{\Delta}$ too. (Also, we have used the fact that $\vartheta(-\vec{z}) = \vartheta(\vec{z})$.) We could determine $\vec{\Delta}$ by arguing backwards from some of the Corollaries of Riemann's theorem, or else we can go back to the proof of II.3.1 and work a little harder on the integrals there. We shall do this although the reader should be warned that the details are such that it is almost impossible not to make mistakes of sign, orientation conventions, etc. The result is that for hyperelliptic C,

(5.4) $$\vec{\Delta} = \Omega\vec{\delta}' + \vec{\delta}'' \mod L$$

where

$$\delta' = \left(\tfrac{1}{2}\,\tfrac{1}{2}\,\cdots\,\tfrac{1}{2}\right) \in \tfrac{1}{2}\mathbb{Z}^g$$

$$\delta'' = \left(\tfrac{g}{2}\,\tfrac{g-1}{2}\,\cdots\cdots\,1\,\tfrac{1}{2}\right) \in \tfrac{1}{2}\mathbb{Z}^g .$$

Proof: Recall the expression for $\vec{\Delta}$ mod L_Ω: let g_k be the indefinite integral of ω_k on $C - \cup A_i' - \cup B_i'$, normalized so that $g_k(\infty) = 0$ (we are extending A_i and B_i by "tails" to get a figure homeomorphic to the one in §II.2). Then

$$\Delta_k \equiv -\frac{\Omega_{kk}}{2} - \int_\infty^{P_1} \omega_k + \sum_{\ell=1}^{g} \int_{A_\ell'^+} g_k\omega_\ell .$$

(In the term $\int_\infty^{P_1} \omega_k$, the path should be taken in $C-\cup A_i'-\cup B_i'$ from ∞ to P_1 considered as the beginning of B_k'.) Firstly, $\omega_\ell = dg_\ell$, so the $\ell = k$ term in the last sum is $\frac{1}{2}\int_{A_k'^+} d(g_k^2)$. This is

$\frac{1}{2}[g_k^2(\text{end of } A_k'^+) - g_k^2(\text{beginning of } A_k'^+)]$. But the end of $A_k'^+$ is the beginning of B_k', and g_k at the beginning of $A_k'^+$ is 1 less than at the end because $\int_{A_k} \omega_k = +1$. So this term is

$$\frac{1}{2}\left[\left(\int_{\infty}^{P_1} \omega_k\right)^2 - \left(\int_{\infty}^{P_1} \omega_k - 1\right)^2\right]$$

$$= \int_{\infty}^{P_1} \omega_k - \frac{1}{2} \quad .$$

Secondly, if $\ell \neq k$, then g_k has the same value at the beginning and end of $A_\ell'^+$ because $\int_{A_\ell} \omega_k = 0$. So the contribution of the tails on $A_\ell'^+$ in

$$\int_{A_\ell'^+} g_k \omega_\ell$$

is zero and we may as well integrate around A_ℓ^+. g_k is evaluated on A_ℓ^+ by paths as follows, missing all A_i, B_i:

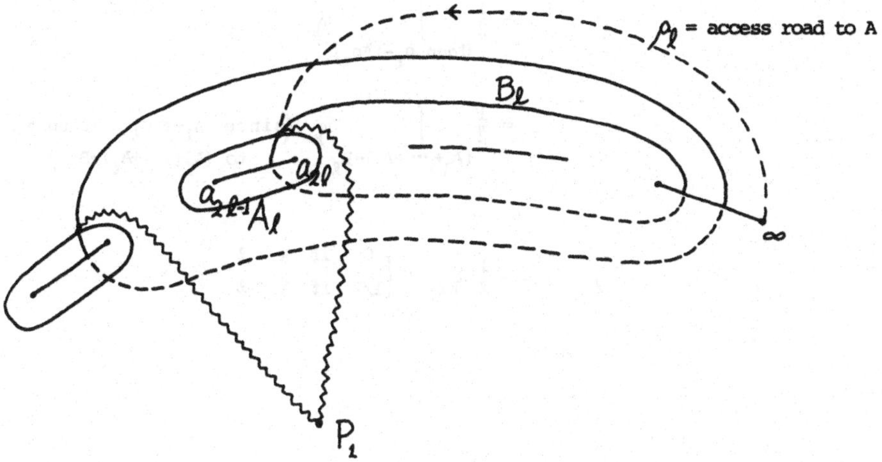

Here we have chosen A_ℓ exactly along the cut between $a_{2\ell-1}, a_{2\ell}$, so that it consists in a path α_ℓ from $a_{2\ell}$ to $a_{2\ell-1}$ and a return along $\iota(\alpha_\ell)$. Now $\iota^*\omega_\ell = -\omega_\ell$, and $\iota(\alpha_\ell)$ is traversed backwards: so

$$\int_{A_\ell^+} g_k\omega_\ell = \int_{\alpha_\ell} (g_k + \iota^* g_k)\omega_\ell .$$

But $g_k+\iota^*g_k$ is constant because $d(g_k+\iota^*g_k) = \omega_k+\iota^*\omega_k = 0$. Thus

$$\begin{aligned}
\int_{\alpha_\ell} (g_k+\iota^*g_k)\omega_\ell &= 2g_k(a_{2\ell}) \int_{\alpha_\ell} \omega_\ell \\
&= g_k(a_{2\ell}) \qquad \left(\text{since } \int_{\alpha_\ell} \omega_\ell = \frac{1}{2}\int_{A_\ell} \omega_\ell = \frac{1}{2}\right) \\
&= \int_{\rho_\ell} \omega_k \\
&= \frac{1}{2} \int_{[\text{loop } \rho_\ell - \iota^*\rho_\ell]} \omega_k \\
&= \frac{1}{2} \int_{(A_1+\cdots+A_\ell)-B_\ell} \omega_k \quad \left(\text{since } \rho_\ell - \iota^*\rho_\ell \text{ is homologous to } (A_1+\cdots+A_\ell)-B_\ell\right) \\
&= -\frac{1}{2}\Omega_{k\ell} + \begin{cases} 0 & \text{if } k > \ell \\ 1/2 & \text{if } k < \ell \end{cases}
\end{aligned}$$

Altogether, this shows

$$\Delta_k \equiv -\frac{1}{2}\sum_{\ell=1}^{g} \Omega_{k\ell} + \begin{cases} 0 & \text{if } g-k+1 \text{ even} \\ 1/2 & \text{if } g-k+1 \text{ odd} \end{cases}$$

which proves (5.4).

Part (2) is just a restatement of Part (1). Before proving (3), we need to tie together the different descriptions we have introduced for the 2-torsion in Pic C. In fact, in §2, we showed that

$$(\text{Pic } C)_2 = \left(\begin{matrix} \text{group of divisor classes } e_T \\ T \subset B, \ \#T \text{ even, mod } e_T = e_{CT} \end{matrix} \right)$$

where

$$e_T = \sum_{P \in T} P - \#T.\infty \ .$$

We also know

$$\begin{aligned} (\text{Pic } C)_2 &\cong \text{2-torsion in } \mathbb{C}^g/L_\Omega \\ &\cong \tfrac{1}{2}L_\Omega/L_\Omega \ . \end{aligned}$$

The link between these is given by

$$e_T \longmapsto I(e_T) = \sum_{P \in T} \int_\infty^P \vec{\omega} \ .$$

We can calculate $I(e_T)$:

<u>Lemma 5.6</u>: a) $I(e_{\{a_{2i-1},a_{2i}\}}) = {}^t(0,\cdots,\overset{i^{th}\text{ place}}{\widehat{\tfrac{1}{2}}},\cdots 0) \bmod L_\Omega$

$$I(e_{\{a_{2i},a_{2i+1},\cdots,a_{2g+1}\}}) = {}^t\left(\frac{\Omega_{1i}}{2},\cdots,\frac{\Omega_{gi}}{2}\right) \bmod L_\Omega$$

b) $I(e_{\{a_{2i-1},\infty\}}) = {}^t\left(\frac{1}{2},\frac{1}{2},\cdots,\frac{1}{2},\overset{i^{th}\text{ place}}{\widehat{0}},\cdots,0\right) + {}^t\left(\frac{\Omega_{1i}}{2},\cdots,\frac{\Omega_{gi}}{2}\right) \bmod L_\Omega$

$$I(e_{\{a_{2i},\infty\}}) = {}^t\left(\frac{1}{2},\frac{1}{2},\cdots,\overset{i^{th}\text{ place}}{\widehat{\tfrac{1}{2}}},0,\cdots,0\right) + {}^t\left(\frac{\Omega_{1i}}{2},\cdots,\frac{\Omega_{gi}}{2}\right) \bmod L_\Omega$$

<u>Proof</u>: The path A_i in the diagram (5.1) above may be moved so that it follows ρ from a_{2i} to a_{2i-1} on one sheet of C, and then goes back on the other sheet:

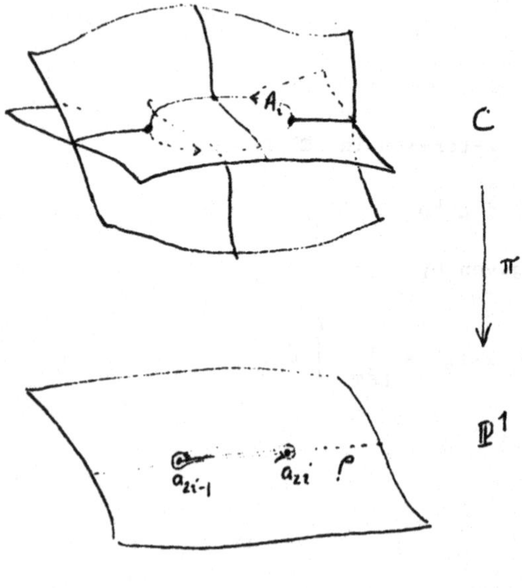

But each ω_k reverses its sign when you switch sheets. As the direction in which A_i is traversed also changes:

$$\frac{1}{2}\int_{A_i} \vec{\omega} = \int_{a_{2i}}^{a_{2i-1}} \vec{\omega}$$

$$= \int_{\infty}^{a_{2i-1}} \vec{\omega} - \int_{\infty}^{a_{2i}} \vec{\omega}$$

$$= I(e_{\{a_{2i-1},a_{2i}\}}) \bmod L_\Omega \quad .$$

(Note: $2\int_{\infty}^{a_{2i}} \vec{\omega} \in L_\Omega$ because $I(2e_T) = I(0) = 0 \in \mathbb{C}^g/L_\Omega$.)

The same argument with B_i shows

$$\frac{1}{2}\int_{B_i} \vec{\omega} = I(e_{\{a_{2i},\cdots,a_{2g+1}\}}) \bmod L_\Omega \quad .$$

This proves (a). (b) follows because of

$$\{a_1,a_2\}\circ\cdots\circ\{a_{2i-3},a_{2i-2}\}\circ\{a_{2i},\cdots,a_{2g+1}\} = C\{a_{2i-1},\infty\}$$

and

$$\{a_1,a_2\}\circ\cdots\circ\{a_{2i-1},a_{2i}\}\circ\{a_{2i},\cdots,a_{2g+1}\} = C\{a_{2i},\infty\}$$

and lemma (2.4). <u>QED</u>

Definition 5.7*:

$$\eta_{2i-1} = \text{the } 2\times g \text{ matrix } \begin{pmatrix} 0\cdots 0 & \overset{i^{\text{th}}\text{ place}}{\frac{1}{2}} & 0\cdots 0 \\ \frac{1}{2}\cdots\frac{1}{2} & 0 & 0\cdots 0 \end{pmatrix} = \begin{pmatrix} {}^t\eta'_{2i-1} \\ {}^t\eta''_{2i-1} \end{pmatrix},$$

$$\eta_{2i} = \text{the } 2\times g \text{ matrix } \begin{pmatrix} 0\cdots 0 & \overset{i^{\text{th}}\text{ place}}{\frac{1}{2}} & 0\cdots 0 \\ \frac{1}{2}\cdots\frac{1}{2} & \frac{1}{2} & 0\cdots 0 \end{pmatrix} = \begin{pmatrix} {}^t\eta'_{2i} \\ {}^t\eta''_{2i} \end{pmatrix}$$

$$\eta_T = \sum_{\substack{a_k \in T \\ a_k \neq \infty}} \eta_k, \qquad \text{for all } T \subset B .$$

Then lemma 5.6 says that

$$\boxed{I(e_T) = \Omega\eta'_T + \eta''_T} \tag{5.8}$$

and the more precise version of part (3) of the theorem states that

$$U^D(a_k) = c_k \left[\frac{\vartheta[\delta+\eta_k]\left(\sum \int_\infty^{P_i} \vec{\omega}\right)}{\vartheta[\delta]\left(\sum \int^{P_i} \vec{\omega}\right)} \right]^2$$

To prove this, note that both sides are meromorphic functions on Jac C with poles only on the irreducible divisor Θ. Suppose we prove that both sides are zero precisely on the translate of Θ

*There is an unfortunate conflict here between conventions for row and column vectors. In Ch. 2, $[{\eta' \atop \eta''}]$ was defined for η', η'' columns of height g. In the 19th century, η', η'' were rows of length g (easier to write!). To make these compatible, we must put a transpose in this definition.

by $\int_{\infty}^{a_k} \vec{\omega}$ and vanish to 2nd order there. It follows that the ratio of the LHS and RHS is finite and non-zero on Jac-Θ, hence it is either 1) a constant, or 2) has a zero on Θ and no poles, or 3) has a pole on Θ and no zeroes. But using the fact that a bounded analytic function on a compact analytic space is constant, applied to the ratio or its inverse, we see that 2) and 3) are impossible.

Consider therefore the zeroes. The RHS has a double zero* on the translate of $V(\vartheta[\delta])$ by $\Omega\eta_k' + \eta_k''$. By our remarks above, this is the translate of Θ by $I(e_{\{k,\infty\}})$, i.e., $\int_{\infty}^{a_k} \vec{\omega}$. As for the LHS, as $D = \sum_1^g P_i$,

$$
\begin{aligned}
U^D(a_k) = 0 &\iff P_i = a_k \quad \text{for some } i \\
&\iff (\sum_1^g P_i) - a_k \equiv \text{(effective divisor of degree } g-1) \\
&\iff \text{(divisor class } \sum_1^g P_i - a_k - (g-1)\infty) \in \Theta \\
&\iff \sum_1^g \int_{\infty}^{P_i} \vec{\omega} \in \text{(translate of } \Theta \text{ by } \int_{\infty}^{a_k} \vec{\omega}).
\end{aligned}
$$

---

*Note that if $\vartheta[\delta]$ vanished to some higher order $r \geq 2$ on Θ, this would contradict Riemann's theorem: because for a general choice of $P_1,\cdots,P_g \in C$, $f(P) = \vartheta(\vec{\Delta} - \sum \int_{\infty}^{P_i} \omega + \int_{\infty}^{P} \vec{\omega})$ vanishes to first order at $P = P_1,\cdots,P_g$.

To check the order of vanishing, go back to the covering:

$$\begin{array}{ccc} (\mathbb{C}^g)_{oo} & \xrightarrow{\text{res } \pi} & Z_o \\ \| & & \| \\ \left(\begin{array}{c}\text{g-tuples } P_1,\cdots,P_g \\ \text{s.t. } P_i \neq P_j, \neq \iota(P_j), \\ \text{if } j \neq i, \neq \infty\end{array}\right) & & \left(\begin{array}{c}\text{open set of } Z \text{ of } U,V \\ \text{such that } U \text{ has } g \\ \text{distinct roots}\end{array}\right) \end{array}$$

The group of permutations acts freely on $(C^g)_{oo}$, so π is an unramified covering map between g-dimensional complex manifolds, i.e., they are locally biholomorphic. Now

$$(\text{res } \pi)^{-1}\left(\begin{array}{c}\text{zeroes of} \\ U^D(a_k)\end{array}\right) = \bigcup_{i=1}^{g} [C\times\cdots\times\overbrace{\{a_k\}}^{i^{th}\text{ place}}\times\cdots\times C]$$

The pull-back of the function $U^D(a_k)$ is $f(P_1,\cdots,P_k) = \prod_{i=1}^{g} (t(P_i)-a_k)$. But the function $t-a_k$ on C vanishes to order 2 at the point $s = 0$, $t = a_k$, i.e., at the point we are calling a_k. So f vanishes to order 2 on $\bigcup_{i=1}^{g} pr_1^{-1}(a_k)$ as required. <u>QED</u>

An interesting restatement of part (3) of the Theorem is

Corollary 5.9: The 2g+2-meromorphic functions

$\left(\frac{\vartheta[\eta_k](z)}{\vartheta[0](z)}\right)^2$, $k \in B$, on Jac C span a vector space V of dimension only g+1. In the projective space $\mathbb{P}(V)$, the individual functions lie on a rational curve D of degree g and on this curve, give a finite set projectively equivalent to B in $\mathbb{P}^1$. In this way, we can reconstruct the hyperelliptic curve C from Jac C and ϑ .

Proof: Part (3) says that, up to a translation in $\vec{z}$, these functions are $D \longmapsto U^D(a_k)$. But

$$U^D(a_k) = \sum_{i=0}^{g} U_i^D \cdot a_k^{g-i} \qquad , \; U_i^D = \text{coefficients of } U^D(t).$$

So the 2g+2 function $U^D(a_k)$ are all constant linear combinations of the g+1 functions $D \longmapsto U_i^D$ (including U_0^D which is the constant function 1).

Taking these U_i^D as a basis of V, the individual functions $U^D(a_k)$ have coordinates in V

$$(a_k^g, a_k^{g-1}, \cdots, a_k, 1) \; .$$

The rational curve D in the theorem is just the locus of points in $\mathbb{P}(V)$ whose homogeneous coordinates in V are:

$$(b^g, b^{g-1}, \cdots, b, 1), \qquad \text{some } b \in \mathbb{C}.$$

Thus b is a coordinate on D and the individual functions $U^D(a_k)$ have coordinates $b = a_k$. Thus the Corollary is just a geometric restatement of Part (3). QED

In §3, we described algebraically the translation invariant vector fields on the variety Z of polynomials U,V,W such that $f-V^2 = U\cdot W$. In analytic coordinates $z_1,\cdots,z_g$ on $\mathbb{C}^g$, the translation invariant vector fields are just $\sum c_i \frac{\partial}{\partial z_i}$, $c_i \in \mathbb{C}$. We can tie these together too. The result is:

<u>Proposition 5.10. Let</u> $\omega_i = \phi_i(t)dt/s$, $\phi_i(t) = e_i t^{g-1} + \cdots\cdots$. <u>Then in the isomorphism</u>

$$\mathbb{C}^g/L_\Omega - V(\vartheta[{}^{\delta'}_{\delta''}]) \cong Z,$$

<u>the vector field</u> D_a <u>on</u> Z <u>corresponds to the vector field</u> $-\sum \phi_i(a) \frac{\partial}{\partial z_i}$ <u>and the vector field</u> D_∞ <u>on</u> Z <u>corresponds to the vector field</u> $-\sum e_i \frac{\partial}{\partial z_i}$.

<u>Proof</u>: Let $D(\varepsilon) = \sum_{i=1}^{g} P_i(\varepsilon)$ represent an integral curve of the vector field D_a. Let $c_i(\varepsilon) = t(P_i(\varepsilon))$, $U^\varepsilon(t) = \prod_{i=1}^{g} (t-c_i(\varepsilon)) = U^{D(\varepsilon)}(t)$, and $V^\varepsilon(t) = V^{D(\varepsilon)}(t)$, so that $s(P_i(\varepsilon)) = V^\varepsilon(c_i(\varepsilon))$. Then

$$\frac{\partial}{\partial \varepsilon} U^\varepsilon(t) = \frac{V^\varepsilon(a)U^\varepsilon(t)-U^\varepsilon(a)V^\varepsilon(t)}{t-a}.$$

The corresponding curve in $\mathbb{C}^g$-space is $\int_{g\cdot\infty}^{D(\varepsilon)} \vec{\omega}$ and we want to prove

$$\frac{\partial}{\partial \varepsilon}\left(\int_{g\cdot\infty}^{D(\varepsilon)} \vec{\omega}\right)\Bigg|_{\varepsilon=0} = -(\phi_1(a),\cdots,\phi_g(a)).$$

Letting $c_i(0) = c_i$, we calculate $\frac{\partial}{\partial\varepsilon}U^\varepsilon(c_i)$ in 2 ways:

$$\left(\frac{\partial}{\partial\varepsilon} U^\varepsilon\right)(c_i) = -\frac{U^\varepsilon(a)V^\varepsilon(c_i)}{c_i-a}$$

and
$$\frac{\partial}{\partial\varepsilon}(U^\varepsilon(c_i)) = \prod_{k\neq i}(c_i-c_k)\left(-\frac{\partial}{\partial\varepsilon}c_i(\varepsilon)\right).$$

Therefore
$$\frac{\partial}{\partial\varepsilon}(c_i(\varepsilon)) = \frac{U^\varepsilon(a)\cdot V^\varepsilon(c_i)}{(c_i-a)\cdot\prod_{k\neq i}(c_i-c_k)}.$$

Letting $t=a$, $s=b$ be the point on C over a, we recall the rational function

$$\frac{U(a)\cdot(s+V(t))+U(t)\cdot(b-V(a))}{U(t)\cdot(t-a)}$$

on C used in §3, which has poles at $P = (a,b)$ and at $P_1,\cdots,P_g$. Take its product with ω_j and use the fact that the sum of its residues at all poles is zero:

$$0 = \sum \operatorname{res}_Q\left(\frac{U(a)\cdot(s+V(t))+U(t)\cdot(b-V(a))}{U(t)\cdot(t-a)}\cdot\frac{\phi_j(t)dt}{s}\right)$$

$$= \operatorname{res}_P\left(\frac{2U(a)\cdot b}{U(a)\cdot(t-a)}\cdot\frac{\phi_j(a)dt}{b}\right) + \sum_i \operatorname{res}_{P_i}\left(\frac{2U(a)V(c_i)}{(t-c_i)\prod_{k\neq i}(c_i-c_k)(c_i-a)}\cdot\frac{\phi_j(c_i)dt}{V(c_i)}\right)$$

$$\left(\text{using } s(P_i) = V(c_i)\right)$$

$$= 2\phi_j(a) + 2\sum_i\left(\frac{\partial}{\partial\varepsilon}(c_i(\varepsilon))\Big|_{\varepsilon=0}\right)\cdot\frac{\phi_j(c_i)}{V(c_i)}.$$

But

$$\frac{\partial}{\partial \varepsilon} \int_{g\cdot\infty}^{D(\varepsilon)} \omega_j \Bigg|_{\varepsilon=0} = \sum_i \frac{\partial}{\partial \varepsilon} \int_{\infty}^{c_i(\varepsilon)} \frac{\phi_j(t)dt}{s} \Bigg|_{\varepsilon=0}$$

$$= \sum_i \frac{\phi_j(c_i)}{V(c_i)} \cdot \frac{\partial}{\partial \varepsilon} (c_i(\varepsilon)) \Bigg|_{\varepsilon=0}$$

$$= -\phi_j(a) \ .$$

The proof for the vector field D_∞ is similar. <u>QED</u>

§6. Theta characteristics and the fundamental Vanishing Property

The appearance of $\vec{\Delta}$ in the main theorem of §5 looks quite mysterious. It appeared as a result of an involved evaluation of the integrals in Riemann's derivation. As in the Appendix to §3, Ch. II, we would like to introduce the concept of theta characteristics in order to give a more intrinsic formulation of (5.3) and clarify the reason for the peculiar looking constant $\vec{\Delta}$. It cannot be eliminated but it can be made to look more natural in this setting.

Recall that theta characteristics on a curve C are divisor classes D such that $2D \equiv K_C$. For hyperelliptic curves, we can describe them as follows:

Proposition 6.1:

i) $K_C \equiv (g-1)L$

ii) Every theta characteristic is of the form

$$f_T \underset{\text{def}}{=\!=} \sum_{P\in T} P + \left(\frac{g-1-\#T}{2}\right)L$$

for some subset $T \subset B$ with $\#T \equiv (g+1)\ (\text{mod } 2)$.

iii) $f_{T_1} \equiv f_{T_2}$ if and only if $T_1 = T_2$ or CT_2, hence the set Σ of theta characteristics is described by:

$$\Sigma \cong \left\{\begin{matrix}\text{set of subsets } T \subset B \\ \#T \equiv (g+1) \text{ mod } 2\end{matrix}\right\} \Big/ \begin{matrix}\text{modulo} \\ T \sim CT\end{matrix}$$

iv) <u>For all such</u> T, $\exists P_1,\cdots,P_{g-1} \in C$ <u>such that</u> $\sum_1^{g-1} P_i \equiv f_T$ <u>if and only if</u> $\#T \neq g+1$, <u>and if</u> $\#T < g+1$

$$\dim \mathcal{L}(f_T) = \dim \mathcal{L}\left(\sum_1^{g-1} P_i\right)$$

$$= \frac{g+1-\#T}{2}$$

(<u>if</u> $\#T > g+1$, <u>replace</u> T <u>by</u> CT to compute $\dim \mathcal{L}(f_T)$).

<u>Proof</u>: In the proof of (5.2), we saw that the divisor of the differential dt/s was just $(2g-2)\infty$, which belongs to the divisor class $(g-1)L$. This proves (i). As for (ii) and (iii), note that

$$2f_T \equiv \sum_{P\in T} 2P + (g-1-\#T)L \equiv (g-1)L$$

hence $f_T \in \Sigma$. But all 2-torsion is representable as divisor classes e_S, and it's immediate that:

(6.2) $$f_T + e_S \equiv f_{T\circ S}\ .$$

Since any 2 theta characteristics differ by 2-torsion, they are all of the form f_T for some T. Moreover

$$f_{T_1} \equiv f_{T_2} \iff e_{T_1\circ T_2} = 0 \quad \text{by (6.2)}$$

$$\iff T_1\circ T_2 = \phi \text{ or } B$$

$$\iff T_1 = T_2 \text{ or } T_1 = CT_2.$$

Finally, to calculate $\mathcal{L}(f_T)$, use

$$f_T \equiv (g-1+\#T)\infty - \sum_{P\in T} P,$$

hence

$$\mathcal{L}(f_T) \cong \begin{pmatrix}\text{space of fcns. } f \text{ with } (g-1+\#T)\text{-fold} \\ \text{pole at } \infty \text{ and zeroes at all } P\in T\end{pmatrix}$$

We assume $\#T \le g+1$, so $(g-1+\#T) \le 2g$. Now functions with 2g-fold poles at ∞ are polynomials in t of degree $\le g$ (s has a $2g+1$-fold pole at ∞). So

$$\mathcal{L}(f_T) \cong \begin{pmatrix}\text{polynomials in } t \text{ of degree} \le (\frac{g-1+\#T}{2}) \\ \text{zero at all } P\in T\end{pmatrix}$$

The dimension of the latter space is $\frac{g+1-\#T}{2}$, hence (iv). <u>QED</u>

Comparing Prop. 6.1 with II.3.10, we come up with a set of canonical isomorphisms as follows:

$$\left\{\begin{matrix}T\subset B\text{: } \#T\equiv g+1\,(2) \\ \text{and } T\sim CT\end{matrix}\right\} \xrightarrow{\sim} \Sigma \xrightarrow{\sim} \left\{\begin{matrix}\text{symmetric translates} \\ \text{of } \Theta \text{ in Jac } C\end{matrix}\right\} \xleftarrow{\sim} \tfrac{1}{2}\mathbb{Z}^{2g}/\mathbb{Z}^{2g}$$

$$T \longmapsto f_T \longmapsto \begin{pmatrix}\text{locus of div. classes} \\ P_1+\cdots+P_{g-1}-f_T\end{pmatrix}$$

$$\text{zeroes of } \vartheta[\eta] \longleftarrow\!\!\!\shortmid\; \eta$$

Thus the symmetric translates of Θ in Jac C can be described combinatorially in 2 ways: by subsets T of B and by $\eta \in \frac{1}{2}\mathbb{Z}^{2g}/\mathbb{Z}^{2g}$. Riemann's theorem tells us how to link these up. The result can be better phrased like this:

Proposition 6.2: Let $U \subset B$ be the set of g+1 branch points $a_1, a_3, \cdots, a_{2g+1}$. In the above correspondences, the following objects correspond to each other:

(a)

$$\begin{bmatrix} \phi \text{ if } g \text{ odd} \\ \{\infty\} \text{ if } g \text{ even} \end{bmatrix} \longleftrightarrow (g-1)\infty \longleftrightarrow \begin{bmatrix} \Theta \text{ itself,} \\ \text{i.e., locus of} \\ \begin{pmatrix} P_1+\cdots+P_{g-1} \\ -(g-1)\infty \end{pmatrix} \end{bmatrix} \longleftrightarrow \delta$$

(b) For all $T \subset B$ such that $\#T \equiv (g+1) \bmod 2$:

$$T \longleftrightarrow \eta_{T \circ U}$$

especially:

$$U \longleftrightarrow 0$$

Proof: (a) is a rephrasing of (5.3) part 1, except for the first description. For this note that

$$g \text{ odd} \Longrightarrow \#(\phi) = 0 \equiv (g+1) \bmod 2 \quad \text{and} \quad f_\phi \equiv (g-1)\infty,$$

$$g \text{ even} \Longrightarrow \#(\{\infty\}) = 1 \equiv (g+1) \bmod 2 \quad \text{and} \quad f_{\{\infty\}} \equiv (g-1)\infty.$$

To check (b), build from (a) as follows: for all $S \subset B$, $\#S$ even,

$$\begin{bmatrix} S \text{ if } g \text{ odd} \\ S\circ\{\infty\} \text{ if } g \text{ even}\end{bmatrix} \longleftrightarrow [e_S+(g-1)\infty] \longleftrightarrow \begin{bmatrix}\text{translate} \\ \text{of}\,\Theta \text{ by} \\ I(e_S)\end{bmatrix} \longleftrightarrow [\delta+\eta_S].$$

If g is odd, $\#U$ is even and one checks

$$\eta_U = \begin{pmatrix}\frac{1}{2} & 0\cdots 0 \\ 0 & 0\cdots 0\end{pmatrix}+\begin{pmatrix}0 & \frac{1}{2}\cdots 0 \\ \frac{1}{2} & 0\cdots 0\end{pmatrix}+\cdots+\begin{pmatrix}0 & 0\cdots 0 & \frac{1}{2} \\ \frac{1}{2} & \frac{1}{2}\cdots\frac{1}{2} & 0\end{pmatrix}+\begin{pmatrix}0 & 0\cdots 0 \\ \frac{1}{2} & \frac{1}{2}\cdots\frac{1}{2}\end{pmatrix} = \delta$$

while if g is even $\#(U\circ\{\infty\})$ is even and

$$\eta_{U\circ\{\infty\}} = \eta_{\{a_2,a_4,\cdots,a_{2g}\}} = \begin{pmatrix}\frac{1}{2} & 0\cdots 0 \\ \frac{1}{2} & 0\cdots 0\end{pmatrix}+\begin{pmatrix}0 & \frac{1}{2}\cdots 0 \\ \frac{1}{2} & \frac{1}{2}\cdots 0\end{pmatrix}+\cdots+\begin{pmatrix}0 & 0\cdots\frac{1}{2} \\ \frac{1}{2} & \frac{1}{2}\cdots\frac{1}{2}\end{pmatrix} = \delta .$$

Letting $T = S$ if g is even, $T = S\circ\{\infty\}$ if g is odd, part (b) follows.

<u>QED</u>

This gives the following "explanation" for $\vec{\Delta}$ and δ: the symmetric translates of Θ are — without <u>any</u> unnecessary choices — naturally parametrized by the divisor classes $\sum$, hence by subsets $T \subset B$, $\#T \equiv g+1\,(2)$. The points of order 2 on Jac C are naturally parametrized by subsets $T \subset B$, $\#T$ even. The theta function, after a lot of non-canonical choices, picks out a particular symmetric Θ, i.e., $\vartheta(z) = 0$. (6.2) shows that in effect all these choices just boil down to fixing a "base point" in the set $\sum$ which is the set U of odd-numbered branch points.

In Ch. II, §3, Appendix, we also noted that Σ came with a natural division into even and odd subsets. We can identify this division in the hyperelliptic case:

Proposition 6.3:

a) $e_2(\eta_{S_1},\eta_{S_2}) = (-1)^{\#(S_1 \cap S_2)}$ for $S_i \subset B$, $\#S_i$ even,

b) $e_*(\eta_{T \circ U}) = (-1)^{(\frac{\#T-g-1}{2})}$ for $T \subset B$, $\#T \equiv (g+1)(2)$,

hence:

c) If $T \subset B$ satisfies $\#T \equiv (g+1) \pmod 2$, f_T is an even element of Σ if and only if $\#T \equiv (g+1) \pmod 4$, odd if and only if $\#T \equiv (g-1) \pmod 4$.

Proof: Check (a) as follows:

Note that

$$\#(S_1 \cap (S_2 \circ S_3)) \equiv \#(S_1 \cap S_2) + \#(S_1 \cap S_3) \pmod 2$$

(see figure 6.4), hence $\#(S_1 \cap S_2) \bmod 2$ is a symmetric bilinear $\mathbb{Z}/2\mathbb{Z}$-valued form on the group of subsets of B. When S_1 and S_2 are generators $\{a_{k_1},\infty\},\{a_{k_2},\infty\}$ of this group, one checks the result directly. This proves (6.3a).

(6.4)

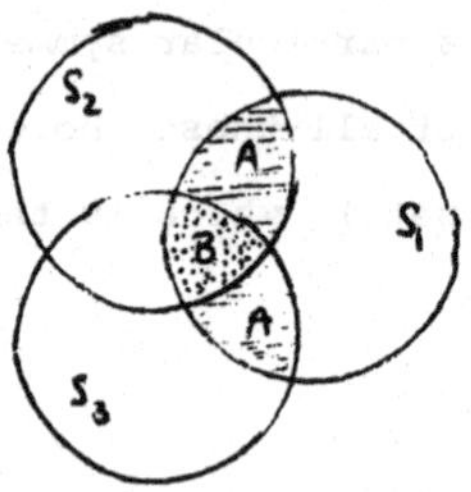

$A = S_1 \cap (S_2 \circ S_3)$

B = points occurring in both $S_1 \cap S_2$ and $S_1 \cap S_3$.

To check (b), recall that

$$\frac{e_*(\alpha+\beta)}{e_*(\alpha)e_*(\beta)} = e_2(\alpha,\beta), \qquad \text{all } \alpha,\beta \in \tfrac{1}{2}\mathbb{Z}^{2g}.$$

Let $e'_*(T) = (-1)^{\frac{\#T-g-1}{2}}$. We check that

$$\frac{e'_*(T\circ S_1\circ S_2)\cdot e'_*(T)}{e'_*(T\circ S_1)e'_*(T\circ S_2)} = (-1)^{\#S_1\cap S_2}$$

for all $S_1,S_2,T\subset B$, $\#S_i$ even, $\#T \equiv g+1(2)$. This is equivalent to:

(6.5) $\#(T\circ S_1\circ S_2) + \#T - \#T\circ S_1 - \#T\circ S_2 \equiv 2\#(S_1\cap S_2)\ (\text{mod } 4)$

Proof by Venn diagram:

(6.6)

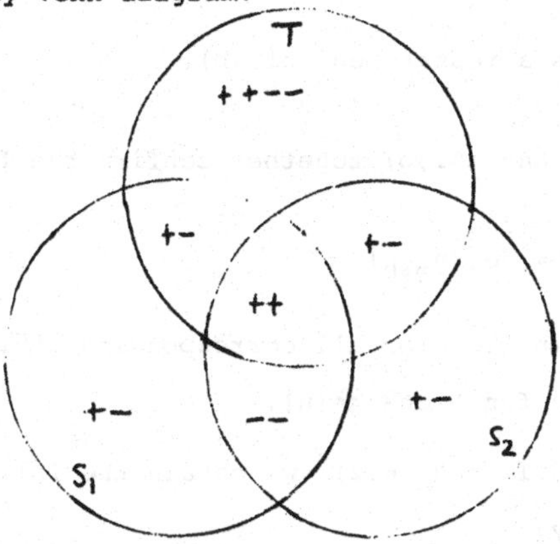

(+ for membership in $T\circ S_1\circ S_2$ or T; - for membership in $T\circ S_1$ or $T\circ S_2$).

Thus in (6.5):

$$\text{LHS} = 2\#(T\cap S_1\cap S_2) - 2\#(S_1\cap S_2\cap CT)$$

$$\equiv 2\#(S_1\cap S_2) \pmod 4 = \text{RHS}.$$

Putting together part (a) and this equality, we find that

$$T \longmapsto e_*'(T\circ U)/e_*(\eta_T)$$

is a homomorphism from the group of even subsets of B to (± 1).
Next check

$$e_*'(U\circ\{a_k,\infty\}) = +1 \quad \text{if } k \text{ is odd}$$

$$= -1 \quad \text{if } k \text{ is even}$$

while $e_*(\eta_k)$ = +1 if k is odd

= -1 if k is even.

This proves (b). (c) is a restatement of (b). QED

Note that (6.1 iv) and (6.3b) together confirm the formula:

$$(-1)^{\dim \mathcal{L}(f_T)} = e_*(\eta_{T\circ U})$$

asserted without proof in II.3 for all corresponding divisor classes D with $2D \equiv K_C$ and theta functions $\vartheta[\eta]$.

Putting together (6.1) and (6.2), we obtain the following very important Corollary:

Corollary 6.7: Let C be a hyperelliptic curve, with branch points B. Describing the topology of C as above, let $U \subset B$ be the (g+1) odd branch points and let Ω be its period matrix. Then for all $S \subset B$, with $\#S$ even, let $I(e_S) \in \text{Jac } C$ be the corresponding 2-division point. Then

$$\vartheta[\eta_S](0,\Omega) = 0 \iff \vartheta(I(e_S),\Omega) = 0 \iff \#(S\circ U) \neq (g+1).$$

Proof: Combine Cor. 3.10 of Ch. II with (6.2) to find:

$$(\vartheta[\eta_S](0,\Omega) = 0) \iff (f_{S\circ U} \equiv P_1+\cdots+P_{g-1} \text{ for some } P_i).$$

Then apply 6.1 iv. QED

The importance of this Corollary is that it provides a lot of pairs $\eta',\eta'' \in \frac{1}{2}\mathbb{Z}^{2g}$ such that for hyperelliptic period matrices Ω,

$$\sum_{n\in\mathbb{Z}^g} \exp(\pi i\,{}^t(n+\eta')\Omega(n+\eta')+2\pi i\,{}^tn\cdot\eta'') = 0 \quad .$$

We know (II.3.14) that for all odd η',η'', i.e., $4\,{}^t\eta'\cdot\eta''$ odd, this vanishes for all Ω because in fact the series vanishes identically. But Cor. 6.7 applies to many even η',η'' as well. We shall see, in fact, that these identities characterize hyperelliptic period matrices. To get some idea of the strength of this vanishing property, it is useful to look a) at low genus and b) to estimate by Stirling's formula, what fraction of the 2-division points are covered by this Corollary for very large genus.

$g = 2$: $\Sigma = \{S \subset \{1,2,3,4,5,6\} \mid \#S = 1, 3, \text{ or } 5\}/(S \sim CS)$

$= [\text{the 6 odd characteristics } \{1\},\{2\},\cdots,\{6\}]$

$\cup \left[\begin{array}{l}\text{the 10 even characteristics} \\ \{1,2,3\},\{1,2,4\},\cdots,\{1,5,6\} \\ (\text{normalizing } S \text{ by assuming } 1 \in S)\end{array}\right]$

$g = 3$: $\Sigma = \{S \subset \{1,2,\cdots,8\} \mid \#S = 0,2,4,6 \text{ or } 8\}/(S \sim CS)$

$= [\text{the one even characteristic } S = \phi \text{ with } \mathcal{L}(f_S) \neq (0)]$

$\cup [\text{the 28 odd characteristics } S = \{i,j\}]$

$\cup [\text{the 35 even characteristics } S = \{1,i,j,k\},\ \mathcal{L}(f_S) = (0)]$

$g = 4$: $\Sigma = \{S \subset \{1,2,\cdots,10\} \mid \#S = 1,3,5,7 \text{ or } 9\}/(S \sim CS)$

$= [\text{the 10 even characteristics } S = \{i\}, \text{ with } \dim \mathcal{L}(f_S) = 2]$

$\cup [\text{the 120 odd theta characteristics } S = \{i,j,k\}]$

$\cup [\text{the 126 even theta char. } S = \{1,i,j,k,\ell\} \text{ with } \mathcal{L}(f_S) = (0)].$

g	Fraction of 2-division pts. which are odd (so that $\vartheta(a,\Omega)=0$ all Ω)	Fraction of 2-division pts. a where $\vartheta(a,\Omega)=0$ Ω hyperelliptic	
2	6/16	6/16	(in dimension 2, hyperelliptic Ω's are an open, dense set)
3	28/64	29/64	
4	120/256	130/256	
5	496/1024	562/1024	
large g	$\sim 1/2$	$\sim\left(1-\frac{2}{\sqrt{\pi}}\frac{1}{\sqrt{g+1}}\right)$	

The last estimate comes from:

$$\#(\{S\subset B \mid \#S=g+1\}/S\sim CS) = \frac{1}{2}\frac{(2g+2)!}{(g+1)!^2}$$

$$\sim \frac{1}{2}\left[\left(\frac{2g+2}{e}\right)^{2g+2}\sqrt{2\pi(2g+2)}\right]\cdot\left[\left(\frac{g+1}{e}\right)^{g+1}\sqrt{2\pi(g+1)}\right]^{-2}$$

$$= 2^{2g}\left[\frac{2}{\sqrt{\pi}}\frac{1}{\sqrt{g+1}}\right] .$$

§7. Frobenius' theta formula

In this section we want to combine Riemann's theta formula (II.6) with the Vanishing Property (6.7) of the last section. An amazing cancellation takes place and we can prove that for hyperelliptic Ω, $\vartheta(\vec{z},\Omega)$ satisfies a much simpler identity discovered in essence by Frobenius*. We shall make many applications of Frobenius' formula. The first of these is to make more explicit the link between the analytic and algebraic theory of the Jacobian by evaluating the constants c_k of Theorem 5.3. The second will be to give explicitly via thetas the solutions of Neumann's dynamical system discussed in §4. Other applications will be given in later sections. Because one of these is to the Theorem characterizing hyperelliptic Ω by the Vanishing Property (6.7), we want to derive Frobenius' theta formula using only this Vanishing and no further aspects of the hyperelliptic situation. Therefore, we assume we are working in the following situation:

1. B = fixed set with $2g+2$ elements
2. $U \subset B$, a fixed subset with $g+1$ elements
3. $\infty \in B-U$ a fixed element
4. $T \longmapsto \eta_T$ an isomorphism:

$$\begin{pmatrix}\text{even subsets of } B\\ \text{modulo } S \sim CS\end{pmatrix} \xrightarrow{\approx} \tfrac{1}{2}\mathbb{Z}^{2g}/\mathbb{Z}^{2g}$$

such that

a) $\eta_{S_1 \circ S_2} = \eta_{S_1} + \eta_{S_2}$

b) $e_2(\eta_{S_1},\eta_{S_2}) = (-1)^{\#S_1 \cap S_2}$

---

* Über die constanten Factoren der Thetareihen, Crelle, 98 (1885); see top formula, p. 249, Collected Works, vol. II.

c) $e_*(\eta_T) = (-1)^{\frac{\#(T \circ U)-g-1}{2}}$

5. $\Omega \in \mathfrak{H}_g$ satisfies $\vartheta[\eta_T](0,\Omega) = 0$ if $\#T\circ U \neq g+1$.

6. We fix $\eta_i \in \frac{1}{2}\mathbb{Z}^{2g}$ for all $i \in B-\infty$ such that $\eta_i \bmod \mathbb{Z}^{2g}$ equals $\eta_{\{i,\infty\}}$, and also let $\eta_\infty = 0$. (This choice affects nothing essentially.)

We shall use the notation

$$\varepsilon_S(k) = \begin{cases} +1 & \text{if } k \in S \\ -1 & \text{if } k \notin S \end{cases}$$

for all $k \in B$, subsets $S \subset B$.

Theorem 7.1 (Generalized Frobenius' theta formula). *In the above situation, for all* $z_i \in \mathbb{C}^g$, $1 \leq i \leq 4$ *such that* $z_1+z_2+z_3+z_4 = 0$, *and for all* $a_i \in \mathbb{Q}^{2g}$, $1 \leq i \leq 4$, *such that* $a_1+a_2+a_3+a_4 = 0$, *then*

$$(F_{ch}) \qquad \sum_{j\in B} \varepsilon_U(j) \cdot \prod_{i=1}^{4} \vartheta[a_i+\eta_j](z_i) = 0$$

or equivalently:

$$(F) \qquad \sum_{j\in B} \varepsilon_U(j) \exp(4\pi i\, {}^t\eta_j' \Omega \eta_j') \prod_{i=1}^{4} \vartheta(z_i+\Omega\eta_j'+\eta_j'') = 0 \quad .$$

<u>Proof</u>: By (R_{ch}), for every $\omega \in \frac{1}{2}\mathbb{Z}^{2g}$

$$2^{-g} \sum_{\lambda\in\frac{1}{2}\mathbb{Z}^{2g}/\mathbb{Z}^{2g}} \exp(-2\pi i\, {}^t\lambda''\cdot(2\omega'))\cdot\vartheta[a_1+2\omega+\lambda](z_1)\cdot\vartheta[a_2+\lambda](z_2)\cdot\vartheta[a_3+\lambda](z_3)\cdot\vartheta[a_4+\lambda](z_4)$$

$$= \vartheta[\omega](0)\cdot\vartheta\left[\frac{a_1+2\omega+a_2-a_3-a_4}{2}\right](z_1+z_2-z_3-z_4)(\cdots\cdot)(\cdots\cdot)$$

or $\vartheta[\omega](0) = 0 \Longrightarrow$

$$0 = \sum_{\lambda\in\frac{1}{2}\mathbb{Z}^{2g}/\mathbb{Z}^{2g}} \exp(4\pi i({}^t\lambda'\cdot\omega''-{}^t\lambda''\cdot\omega'))\cdot\prod_{i=1}^{4}\vartheta[a_i+\lambda](z_i),$$

since

$$\vartheta[a_1+2\omega+\lambda](z_1) = e^{2\pi i\,{}^t(a_1'+\lambda')\cdot 2\omega''}\vartheta[a_1+\lambda](z_1).$$

Therefore, $\forall\, T\subset B$, $\#T$ even, $\#T\circ U \neq (g+1)$,

$$0 = \sum_{\substack{S\subset B,\,\#S \text{ even}\\ S\sim CS}} (-1)^{\#S\cap T}\cdot\prod_{i=1}^{4}\vartheta[a_i+\eta_S](z_i).$$

Thus, for any coefficients c_T,

$$(7.2)\qquad 0 = \sum_{\substack{S\subset B\\ \#S \text{ even}\\ \text{mod } S\sim CS}}\left[\sum_{\substack{T\subset B\\ \#T\text{ even}\\ \#(T\circ U)\neq g+1}} c_T\,(-1)^{\#S\cap T}\right]\prod_{i=1}^{4}\vartheta[a_i+\eta_S](z_i).$$

What we must do is to choose the c_T's so that "most" but not all of the terms in brackets vanish! For this, we resort to a combinatorial lemma:

Lemma 7.3. For all $S \subset B$, $\#S$ even,

$$\sum_{\substack{T\subset B \\ \infty \in T \\ \#T \equiv (g+1) \bmod 2}} \frac{\#T-(g+1)}{2} \cdot (-1)^{\#S\cap T} = \begin{cases} 0 & \text{if } S \neq \phi, \{\infty,k\}, B-\{\infty,k\}, B \\ 2^{g-2} & \text{if either } S = \phi, \{\infty,k\} \text{ or } \\ & S = B-\{\infty,k\}, B \text{ and } g \text{ is odd} \\ -2^{2g-2} & \text{if } S = B, B-\{\infty,k\} \text{ and } g \text{ is even.} \end{cases}$$

Proof: We note first the following points:

a) for all finite non-empty sets R,

$$\#\begin{pmatrix}\text{subsets } T\subset R \\ \text{s.t. } \#T \text{ even}\end{pmatrix} = \#\begin{pmatrix}\text{subsets } T\subset R \\ \text{s.t. } \#T \text{ odd}\end{pmatrix} = 2^{\#R-1} .$$

In fact, the subsets $T \subset R$ form a group under $\circ$ and the even subsets are a subgroup of index 2.

b) for all finite sets R with at least 2 elements,

$$\sum_{\begin{pmatrix} T\subset R \\ \#T \text{ even}\end{pmatrix}} (\#T) = \sum_{\begin{pmatrix} T\subset R \\ \#T \text{ odd}\end{pmatrix}} (\#T) = (\#R)\cdot 2^{\#R-2} .$$

In fact, the first sum here is the cardinality of the set of pairs (i,S), where $i \in R$ and S is an odd subset of $R-\{i\}$, and we count this by (a). The second sum is the same except that S is an even subset of $R-\{i\}$.

Given these facts, we can easily work out the sum of the lemma. Note that it is invariant, up to the sign $(-1)^{\#T} = (-1)^{g+1}$, under $S \longmapsto CS = B-S$, so we may assume $\infty \in S$. We then have

$$\sum_{\substack{T \subset B \\ \infty \in T \\ \#T \equiv (g+1) \bmod 2}} \frac{\#T-(g+1)}{2} \cdot (-1)^{\#S\cap T} = \sum_{\substack{\infty \in T_1 \subset S \\ T_2 \subset CS \\ \#T_1+\#T_2 \equiv g+1}} \frac{\#T_1+\#T_2-(g+1)}{2} \cdot (-1)^{\#T_1}$$

$$= \sum_{\infty \in T_1 \subset S} \frac{(-1)^{\#T_1}}{2} \left\{ \sum_{\substack{T_2 \subset CS \\ \#T_2 \equiv (g+1-\#T_1)}} [\#T_2 + (\#T_1-(g+1))] \right\}$$

If $\#CS \geq 2$ and $\#S \geq 3$, then

$$= \sum_{\infty \in T_1 \subset S} \frac{(-1)^{\#T_1}}{2} \left\{ \#CS \cdot 2^{\#CS-2} + [\#T_1-(g+1)]2^{\#CS-1} \right\}$$

$$= 2^{\#CS-3} \left\{ \sum_{\substack{\infty \in T_1 \subset S \\ \#T_1 \text{ even}}} \#CS+2[\#T_1-(g+1)] - \sum_{\substack{\infty \in T_1 \subset S \\ \#T_1 \text{ odd}}} \#CS+2[\#T_1-(g+1)] \right\}$$

$= 0$, using b) again.

If either $\#CS \leq 1$ or $\#S \leq 2$, we must have $S = \{\infty,k\}$ or $S = B$; in the first case we compute directly:

$$2^{\#CS-3}\left\{ \sum_{\substack{\infty \in T_1 \subset S \\ \#T_1 \text{ even}}} \#CS + 2[\#T_1-(g+1)] - \sum_{\substack{\infty \in T_1 \subset S \\ \#T_1 \text{ odd}}} \#CS+2[\#T_1-(g+1)] \right\} =$$

$$= 2^{2g-3}\left(2g+2[2-(g+1)] - 2g - 2[1-(g+1)]\right) = 2^{2g-2} ;$$

in the second case,

$$\sum_{\substack{\infty \in T_1 \subset S \\ \#T_1 \equiv g+1}} \frac{\#T_1-(g+1)}{2} (-1)^{\#T_1} = \frac{(-1)^{g+1}}{2}[(2g+1)2^{2g+1-3} - (g+1)2^{2g+1-2}]$$

$$= (-1)^{g+1} \cdot 2^{2g-2}.$$ QED

To apply the lemma, note that

$$\#S \cap (T \circ U) \equiv \#(S \cap T) + \#(S \cap U) \mod 2.$$

In formula (7.2), set

$$c_T = \begin{cases} \dfrac{\#(T \circ U)-(g+1)}{2} & \text{if } \infty \in T \circ U \\ 0 & \text{if not} \end{cases}$$

Then by the lemma,

$$0 = \sum_{\substack{S \subset B \\ \#S \text{ even} \\ \text{mod } S \sim CS}} \left[\sum_{\substack{T \subset B \\ \#T \text{ even}}} \frac{\#(T \circ U)-(g+1)}{2} \cdot (-1)^{\#S \cap (T \circ U)} \cdot (-1)^{\#(S \cap U)} \right] \prod_{i=1}^{4} \vartheta[a_i + \eta_S](z_i)$$

$$= \sum_{\substack{S=\{\infty\}\circ\{k\} \\ k \in B}} 2^{2g-2} (-1)^{\#S \cap U} \cdot \prod_{i=1}^{4} \vartheta[a_i+\eta_S](z_i).$$ QED

Corollary 7.4. Let $T \subset B$ have $g+2$ elements and let $S = T \circ U \circ \{\infty\}$, so $\#S$ even. Then

$$\sum_{j \in T} \varepsilon_\infty(j) \exp(4\pi i\, {}^t\eta_S'' \cdot \eta_j')\, \vartheta[\eta_S+\eta_j](0)^2 \cdot \vartheta[\eta_j](z)^2 = 0.$$

Proof. In F_{ch}, take

$$z_1 = z_2 = 0, \quad z_3 = z, \quad z_4 = -z$$

$$a_1 = \eta_S,\ a_2 = -\eta_S, \quad a_3 = a_4 = 0.$$

Then

$$0 = \sum_{j \in B} \varepsilon_U(j)\, \vartheta[\eta_S+\eta_j](0) \cdot \vartheta[-\eta_S+\eta_j](0)\, \vartheta[\eta_j](z) \cdot \vartheta[\eta_j](-z);$$

now for any $\lambda \in \mathbb{Z}^{2g}$

$$\vartheta[\alpha+\lambda](z) = \exp(2\pi i\, {}^t\alpha' \cdot \lambda'')\, \vartheta[\alpha](z),$$

so

$$\vartheta[-\eta_S+\eta_j](0) = \vartheta[\eta_S+\eta_j-2\eta_S](0) = \exp(-4\pi i\, {}^t(\eta_S'+\eta_j') \cdot \eta_S'')\, \vartheta[\delta_S+\eta_j](0),$$

and

$$\vartheta[\eta_j](-z) = e_*(\eta_j)\, \vartheta[\eta_j](z).$$

But

$$e_*(\eta_j) = (-1)^{\left(\frac{\#U \circ \{j\} \circ \{\infty\} - g - 1}{2}\right)} = \varepsilon_{U \circ \{\infty\}}(j),$$

so putting all this together, we have the formula (7.4). QED

Corollary 7.5: $$\sum_{j\in U}\left(\frac{\vartheta[\eta_j](0)\cdot\vartheta[\eta_j](z)}{\vartheta[0](0)\cdot\vartheta[0](z)}\right)^2 = 1 .$$

Proof: In (7.4), set $T = U \cup \{\infty\}$, hence $S = \phi$.

We now apply Frobenius' identity to refine (5.3) above:

Theorem 7.6: As in (5.3) consider the map

$$\begin{array}{ccc} \text{Jac } C - \Theta & \longrightarrow & \left(\begin{array}{c}\text{space of monic polyn.}\\ U(t) \text{ of degree } g\end{array}\right) \cong \mathbb{C}^g \\ \cup\!\!\!| & & \cup\!\!\!| \\ D & \longmapsto & U^D(t) \end{array}$$

Then for all finite branch points a_k, $1 \leq k \leq 2g+1$, and for all $V \subset \{1,2,\cdots,2g+1\}$ such that $\#V = g+1$, $k \in V$

$$U^D(a_k) = \pm \prod_{\substack{i\in V\\ i\neq k}} (a_k-a_i)\cdot\left(\frac{\vartheta[\eta_{U\circ V}+\eta_k](0)\cdot\vartheta[\delta+\eta_k](z)}{\vartheta[\eta_{U\circ V}](0)\cdot\vartheta[\delta](z)}\right)^2$$

where $\vec{z} = \int_D \vec{\omega}$ and the sign is given by

$$(-1)^{4\,{}^t\delta'\cdot\eta_k''}\cdot(-1)^{4\cdot{}^t\eta_{U\circ V}''\cdot\eta_k'} .$$

Proof: We make a partial fraction expansion:

$$\frac{U^D(t)}{\prod_{k\in V}(t-a_k)} = \sum_{k\in V}\frac{\lambda^D_{k,V}}{t-a_k} .$$

Then $U^D(t)$ is monic but otherwise arbitrary, so $\sum_k \lambda^D_{k,V} = 1$ but otherwise the $\lambda^D_{k,V}$ are arbitrary. In particular, if $\sum d_k\lambda^D_{k,V} = 1$, for all D, then $d_k = 1$, all k. Now by (5.3),

$$\lambda_{k,V}^{D} = \frac{U^D(a_k)}{\prod_{\substack{i\in V\\ i\neq k}}(a_k-a_i)} = \frac{c_k}{\prod_{\substack{i\in V\\ i\neq k}}(a_k-a_i)} \cdot \left(\frac{\vartheta[\delta+\eta_k](z)}{\vartheta[\delta](z)}\right)^2$$

On the other hand, by (7.4), with $T = V \cup (\infty)$, $S = U\circ V$,

$$(*) \quad 1 = \sum_{k\in V} \exp(4\pi i\, {}^t\eta''_{U\circ V}\cdot\eta'_k) \left(\frac{\vartheta[\eta_{U\circ V}+\eta_k](0)\cdot\vartheta[\eta_k](z)}{\vartheta[\eta_{U\circ V}[(0)\cdot\vartheta[0](z)}\right)^2$$

Using the definition of ϑ-functions with characteristic, it follows:

$$\left(\frac{\vartheta[\delta+\eta_k](z)}{\vartheta[\delta](z)}\right)^2 = \exp(4\pi i\, {}^t\delta'\cdot\eta''_k) \left(\frac{\vartheta[\eta_k](z+\Omega\delta'+\delta'')}{\vartheta[0]\ \ (z+\Omega\delta'+\delta'')}\right)^2 .$$

Since z is arbitrary in (*), we can replace it by $z+\Omega\delta'+\delta''$ and find:

$$1 = \sum_{k\in V} \exp(4\pi i\, {}^t\eta''_{U\circ V}\cdot\eta'_k + 4\pi i\, {}^t\delta'\cdot\eta''_k) \left(\frac{\vartheta[\eta_{U\circ V}+\eta_k](0)}{\vartheta[\eta_{U\circ V}](0)}\right)^2 \cdot \left(\frac{\vartheta[\delta+\eta_k](z)}{\vartheta[\delta](z)}\right)^2$$

$$= \sum_{k\in V} \exp(4\pi i\cdots)\cdot\left(\frac{\vartheta[\eta_{U\circ V}+\eta_k](0)}{\vartheta[\eta_{U\circ V}](0)}\right)^2 \cdot \frac{\prod_{i\in V, i\neq k}(a_k-a_i)}{c_k}\cdot\lambda_{k,V}^{D}$$

for <u>all</u> D. Thus the coefficients of $\lambda_{k,V}^{D}$ are all 1, hence

$$c_k = \exp(4\pi i\cdots)\cdot\prod_{i\in V, i\neq k}(a_k-a_i)\cdot\left(\frac{\vartheta[\eta_{U\circ V}+\eta_k](0)}{\vartheta[\eta_{U\circ V}](0)}\right)^2 .$$

<u>QED</u>

A second application is to the explicit solutions $(x_k(t), y_k(t))$ of Neumann's system of differential equations,

$$\begin{aligned} \dot{x}_k &= y_k \\ \dot{y}_k &= -a_k x_k + x_k\left(\sum a_i x_i^2 - \sum y_i^2\right) \end{aligned}$$

where $a_1 < \cdots < a_{g+1}$ are fixed real numbers and $\sum x_k^2 = 1$ and $\sum x_k y_k = 0$. We saw that

$$F_k(x,y) = x_k^2 + \sum_{\ell \neq k} \frac{(x_k y_\ell - x_\ell y_k)^2}{a_k - a_\ell}$$

are integrals of this motion and set up the following maps:

$$\begin{array}{ccccc}
T_{\mathbb{C}}(S^{g+1}) \supset \begin{pmatrix} \text{subvariety} \\ F_1(x,y)=c_1 \\ \cdots\cdots \\ F_{g+1}(x,y)=c_{g+1} \end{pmatrix} & & & & \\
\Big\downarrow \pi & & & & \\
\mathbb{C}^{3g+1} \supset \begin{pmatrix} \text{space of polyn.} \\ U(t), V(t), W(t) \\ \text{s.t.} \\ f - V^2 = U \cdot W \end{pmatrix} & \xleftarrow{\approx} & (\text{Jac } C - \Theta) & \xrightarrow[I]{\approx} & \mathbb{C}^g/L_\Omega - \begin{pmatrix} \text{zeroes of} \\ \vartheta[\delta](z,\Omega) \end{pmatrix}
\end{array}$$

Here $T_{\mathbb{C}}(S^{g+1})$ is the affine variety of x,y such that $\sum x_i^2 = 1$, $\sum x_i y_i = 0$, $\pi(x,y) = (U,V,W)$ where

$$\begin{aligned} U(t) &= f_1(t) \sum \frac{x_k^2}{t-a_k} \\ V(t) &= f_1(t)\sqrt{-1} \sum \frac{x_k y_k}{t-a_k} \\ W(t) &= f_1(t) \cdot \left(\sum \frac{y_k^2}{t-a_k} + 1\right) \end{aligned}$$

$f_1(t) = \prod_{k=1}^{g+1} (t-a_k)$, $f_2(t) = f_1(t) \cdot \sum \frac{c_k}{t-a_k}$, $f(t) = f_1(t) f_2(t)$. We shall assume for simplicity that the constants c_k are all chosen to be positive. The other cases may be treated quite similarly. This means that $\operatorname{sign} f_2(a_k) = (-1)^{g+1+k}$, hence the zeroes $b_1, \cdots, b_g$ of f_2 satisfy:

$$a_1 < b_1 < a_2 < b_2 < \cdots\cdots < a_g < b_g < a_{g+1}$$

Graph of f:

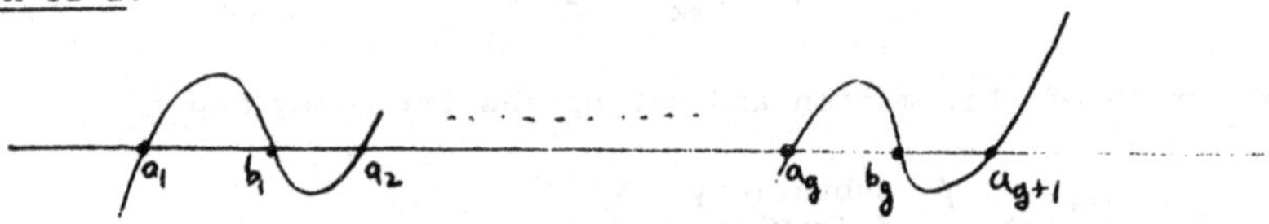

We assume that the cycles A_i, B_i on the curve C given by $s^2 = f(t)$ are chosen as in §5, with respect to the linear ordering of the branch points on the real axis.

Neumann's equations are the equations given by the Hamiltonian vector field X_H on $T_{\mathbb{C}}(S^{g+1})$, which is tangent to the above subvariety. We have seen that

$$\pi_* X_H = -2\sqrt{-1} \cdot D_\infty$$

and that the vector field D_∞ on Jac C is given by

$$-\sum e_i (\partial/\partial z_i)$$

on $\mathbb{C}^g$, where if ω_i are the normalized 1-forms on C, $\omega_i = \phi_i(t)dt/s$ and $\phi_i(t) = e_i t^{g-1} + \cdots$. Therefore the solutions $(x_k(t), y_k(t))$, $t \in \mathbb{R}$, of Neumann's equations project to curves on Jac C-Θ, which lift to the straight lines:

$$\vec{z}_0 + 2\sqrt{-1}\, t\, \vec{e}, \qquad t \in \mathbb{R}, \quad \vec{e} = (e_1,\cdots,e_g),$$

in $\mathbb{C}^g$. Moreover:

$$x_k^2 = \prod_{\ell\neq k} (a_k-a_\ell)^{-1}\cdot U(a_k)$$

$$= \pm\left(\frac{\vartheta[\eta_{2k-1}](0)\cdot\vartheta[\delta+\eta_{2k-1}](z)}{\vartheta[0](0)\cdot\vartheta[\delta](z)}\right)^2$$

by (7.6), where in (7.6) we choose $V = U = \{1,3,\cdots,2g+1\}$ (i.e., corresponding to the branch points $a_1,\cdots,a_{g+1}$), and $\vec{z} = \int_{g\infty}^{D} \vec{\omega}$, D = (divisor defined by U(t) = 0, s = V(t)). The η_{2k-1} appears because a_k is the $(2k-1)^{st}$ branch point in the linear ordering. The sign becomes +1 if we put the characteristic δ back into a translation by $\vec{\Delta}$:

$$x_k^2 = +\left(\frac{\vartheta[\eta_{2k-1}](0)\cdot\vartheta[\eta_{2k-1}](\vec{z}-\vec{\Delta})}{\vartheta[0](0)\cdot\vartheta[0](\vec{z}-\vec{\Delta})}\right)^2$$

(see proof of (7.6)). Now note that whereas $\left(\vartheta[\eta_{2k-1}](z)/\vartheta[0](z)\right)^2$ is periodic with respect to L_Ω, $\vartheta[\eta_{2k-1}](z)/\vartheta[0](z)$ is not. In fact,

$$\frac{\vartheta[\eta_{2k-1}](z+\Omega n+m)}{\vartheta[0](z+\Omega n+m)} = (-1)^{m_k+n_1+\cdots+n_{k-1}}\cdot\frac{\vartheta[\eta_{2k-1}](z)}{\vartheta[0](z)}$$

(Ch. II, 1 and Def. (5.7)). Thus let L'_Ω be the sublattice in L_Ω

of index 2^{g+1} defined by

$$L'_\Omega = \left\{ \Omega n+m \;\middle|\; \begin{array}{l} n,m \in \mathbb{Z}^g \text{ and } m_1, m_2+n_1, \cdots \\ \cdots, m_g+n_1+\cdots+n_{g-1}, n_1+\cdots+n_g \text{ even} \end{array} \right\}$$

These ratios are L'_Ω-periodic. So if we consider the torus $\mathbb{C}^g/L'_\Omega$ which covers $\mathbb{C}^g/L_\Omega$, it follows that we can complete the previous diagram as follows:

$$\begin{array}{ccccc}
T_{\mathbb{C}}(S^{g+1}) \supset & \begin{pmatrix} \text{subvariety} \\ F_1(x,y)=c_1 \\ \cdots\cdots\cdots \\ F_{g+1}(x,y)=c_{g+1} \end{pmatrix} & \xleftarrow{\ \approx\ } & \mathbb{C}^g/L'_\Omega - \begin{pmatrix} \text{zeroes of} \\ \vartheta[\delta](z,\Omega) \end{pmatrix} \\
& \downarrow & & \downarrow \\
\mathbb{C}^{3g+1} \supset & \begin{pmatrix} \text{space of polyn.} \\ U(t),V(t),W(t) \\ \text{s.t.} \\ f-V^2 = UW \end{pmatrix} & \xleftarrow{\ \approx\ } & \mathbb{C}^g/L_\Omega - \begin{pmatrix} \text{zeroes of} \\ \vartheta[\delta](z,\Omega) \end{pmatrix}
\end{array}$$

if we <u>define</u> the upper arrow by

$$x_k = \frac{\vartheta[\eta_{2k-1}](0)\cdot\vartheta[\eta_{2k-1}](\vec{z}-\vec{\Delta})}{\vartheta[0](0)\cdot\vartheta[0](\vec{z}-\vec{\Delta})}$$

Note that the action of the group $L_\Omega/L'_\Omega \cong (\mathbb{Z}/2\mathbb{Z})^{g+1}$ on $\mathbb{C}^g/L'_\Omega$ -(zeroes of $\vartheta[\delta]$) corresponds to the action of the elementary 2-group $(x_k,y_k) \longmapsto (\varepsilon_k x_k, \varepsilon_k y_k)$, $\varepsilon_1,\cdots,\varepsilon_{g+1} \in \{\pm 1\}$ on $T_{\mathbb{C}}(S^{g+1})$. Now the liftings of straight lines in $\mathbb{C}^g/L_\Omega$ are straight lines in $\mathbb{C}^g/L'_\Omega$ so the solution to Neumann's equations are:

$$x_k(t) = \frac{\vartheta[\eta_{2k-1}](0)\cdot\vartheta[\eta_{2k-1}](\vec{z}_0-\vec{\Delta}+2\sqrt{-1}\,t\,\vec{e})}{\vartheta[0](0)\;\cdot\vartheta[0](\vec{z}_0-\vec{\Delta}+2\sqrt{-1}\,t\,\vec{e})},\; t \in \mathbb{R}.$$

Finally, if we want $x_k(t)$ to be <u>real</u>, we have seen that this means that the divisor D given by $U(t) = 0$, $s = V(t)$ should consist in g points $(P_1,\cdots,P_g)$ with $b_i \le t(P_i) \le a_{i+1}$, $s(P_i)$ pure imaginary. In this case

$$\vec{z}_0-\vec{\Delta} = \sum_{i=1}^{g}\int_\infty^{P_i}\vec{\omega} - \sum_{i=1}^{g}\int_\infty^{(b_i,0)}\vec{\omega} = \sum_{i=1}^{g}\int_{(b_i,0)}^{P_i}\vec{\omega}\,.$$

§8. Thomae's formula and moduli of hyperelliptic curves

As a consequence of the formula expressing the polynomial $U^D(t)$ in terms of theta functions, we can directly relate the cross-ratios of the branch points a_i to the "theta-constants" $\vartheta[\eta](0,\Omega)$. This result goes back to Thomae: Beitrag zur bestimmung von $\vartheta(0,\cdots,0)$ durch die Klassenmoduln algebraischer Funktionen, Crelle, 71 (1870). We claim:

Theorem 8.1. For all sets of branch points $B = \{a_1,\cdots,a_{2g+1},\infty\}$, there is a constant c such that for all $S \subset B-\infty$, $\#S$ even,

$$\vartheta[\eta_S](0)^4 = \begin{cases} 0 & \text{if } \#S\circ U \neq g+1 \\ c\cdot(-1)^{\#S\cap U} \cdot \prod\limits_{\substack{i\in S\circ U \\ j\in B-S\circ U-\infty}} (a_i-a_j)^{-1} & \text{if } \#S\circ U = g+1 \end{cases}$$

The result looks more natural if we don't distinguish one branch point by putting it at infinity. Let $B = \{a_1,\cdots,a_{2g+2}\}$ be all finite, put the a_i's on a simple closed curve in this order and choose A_i,B_i as before. Then for all $S \subset B$, $\#S$ even:

$$(8.2) \qquad \vartheta[\eta_S](0)^4 = \begin{cases} 0 & \text{if } \#S\circ U \neq g+1 \\ c\cdot(-1)^{\#S\cap U} \prod\limits_{\substack{i\in S\circ U \\ j\notin S\circ U}} (a_i-a_j)^{-1} & \text{if } \#S\circ U = g+1 \end{cases}$$

In fact Thomae evaluated c too. The answer is:

$$\text{(8.3)}\qquad \vartheta[\eta_S](0)^4 = \pm(\det\ \sigma)^{-2}\cdot \prod_{\substack{i<j\\ i,j\in S\circ U}} (a_i-a_j)\cdot \prod_{\substack{i<j\\ i,j\notin S\circ U}} (a_i-a_j)$$

where

$$2\pi\sqrt{-1}\,.\,\omega_i = \frac{\left(\sum_{j=1}^{g} \sigma_{ij}x^{j-1}\right)dx}{y}\ .$$

For a proof of this, see Fay, <u>op. cit.</u>, p. 46.

We can deduce (8.2) from (8.1) by making a substitution

$$a_i = \frac{Aa_i'+B}{Ca_i'+D} \qquad 1 \le i \le 2g+1$$

$$\infty = \frac{Aa_{2g+2}'+B}{Ca_{2g+2}'+D}, \qquad \text{or} \quad a_{2g+2}' = -\frac{D}{C}\ .$$

The numbers $\vartheta[\eta_S](0)$ are not affected but the RHS changes to

$$c\cdot(-1)^{\#S\cap U} \prod_{\substack{i\in S\circ U\\ j\in B-S\circ U-\infty}} \left\{\frac{a_i'-a_j'}{(Ca_i'+D)(Ca_j'+D)}\right\}^{-1}$$

$$= \left(c\cdot \prod_{i=1}^{2g+2} (Ca_i'+D)^{g+1}\right)\cdot(-1)^{\#S\cap U} \prod_{\substack{i\in S\circ U\\ j\in B-S\circ U-\infty}} (a_i'-a_j')^{-1}\cdot \prod_{i\in S\circ U} (Ca_i'+D)^{-1}$$

$$= \left(cC^{-g-1}\cdot\prod(Ca_i'+D)^{g+1}\right)\cdot(-1)^{\#S\cap U}\ .\ \prod_{\substack{i\in S\circ U\\ j\notin S\circ U}} (a_i'-a_j')^{-1}.$$

To prove (8.1), we substitute $D = \sum_{\substack{1\le i\le 2g+1 \\ i\notin V}} a_i - g\cdot\infty$ into 7.6.

Then

$$U^D(t) = \prod_{i\in V'} (t-a_i)$$

where $V' = \{1,2,\cdots,2g+1\}-V$, and

$$U^D(a_k) = \prod_{i\in V'} (a_k-a_i).$$

But $D \equiv e_V$ if g is odd, $D \equiv e_{V\cup\{\infty\}}$ if g is even. So

$$D \equiv e_{U\circ V} + \begin{cases} e_U & g \text{ odd} \\ e_{U\circ\{\infty\}} & g \text{ even} \end{cases}$$

$$I(D) = I(e_{U\circ V}) + \begin{cases} I(e_U) & g \text{ odd} \\ I(e_{U\circ\{\infty\}}) & g \text{ even} \end{cases}$$

$$= (\Omega\eta'_{U\circ V}+\eta''_{U\circ V}) + (\Omega\delta'+\delta'').$$

Therefore replacing the argument by a theta characteristic:

$$\left(\frac{\vartheta[\delta+\eta_k](I(D))}{\vartheta[\delta](I(D))}\right)^2 = \exp(4\pi i\,{}^t(\delta'+\eta'_{U\circ V})\cdot\eta''_k)\left(\frac{\vartheta[\eta_{U\circ V}+\eta_k](0)}{\vartheta[\eta_{U\circ V}](0)}\right)^2$$

hence (7.6) reads:

$$(8.4)\qquad \frac{\prod_{i\in V'}(a_k-a_i)}{\prod_{i\in V'-\{k\}}(a_k-a_i)} = (-1)^{4({}^t\eta'_{U\circ V}\cdot\eta''_k-{}^t\eta''_{U\circ V}\cdot\eta'_k)}\cdot\frac{\vartheta[\eta_{U\circ V}+\eta_k](0)^4}{\vartheta[\eta_{U\circ V}](0)^4}$$

By (6.3), the sign is

$$(-1)^{\#(U\circ V)\cap\{k,\infty\}}\ .$$

Since $k \in V$, this is +1 if $k \in U$, -1 if $k \notin U$. Now in (8.1), we may choose c to make the formula correct for $S = \phi$, and then prove it for any S by decreasing induction on the number of elements in $(S\circ U)\cap U$: if $(S\circ U) = U$, then $S = \phi$ and we are done. Formula (8.4) is just the ratio of the 2 cases of (8.1):

$$S = U\circ V \quad\text{and}\quad S = C(U\circ(V-k+\infty)).$$

This is straightforward although somewhat painstaking to verify. Therefore applying (8.4) twice, we obtain the ratio of (8.1) for

$$S = U\circ V, \quad\text{and}\quad S = U\circ(V-k+\ell), \quad\text{if}\quad k \in V,\ \ell \notin V.$$

For these $S\circ U$ is respectively V and $V-k+\ell$, hence, step by step, we can move from the formula in the case $S\circ U = U$ to $S\circ U =$ (any V with $\#V = g+1$). QED

We now introduce a moduli space in order to formulate our results more geometrically:

$$(8.5)\qquad \mathcal{H}_g^{(2)} = \left\{\begin{array}{l}\text{set of pairs } (C,\phi),\ C \text{ a hyperelliptic} \\ \text{curve, } \phi: \{1,2,\cdots,2g+1,\infty\} \xrightarrow{\sim} B \text{ a} \\ \text{bijection where } B \subset C \text{ are the branch} \\ \text{points of } \quad \pi: C \longrightarrow \mathbb{P}^1\end{array}\right\} \Big/ \begin{array}{l}\text{mod} \\ \text{isomorphism.}\end{array}$$

We are merely defining $\mathcal{H}_g^{(2)}$ as a set here, the set of isomorphism classes of hyperelliptic curves with "marked" branch points. However because the image of the branch points in $\mathbb{P}^1$ determine the curve, $\mathcal{H}_g^{(2)}$ can be described equivalently as:

$$(8.6)\qquad \mathcal{H}_g^{(2)} \cong \left\{\begin{array}{l}\text{set of sequences } P_1 P_2,\cdots,P_{2g+1},P_\infty \\ \text{of distinct points of } \mathbb{P}^1\end{array}\right\} \Big/ \begin{array}{c}\text{mod projective} \\ \text{equivalence} \\ PGL(2,\mathbb{C})\end{array}$$

Since we can normalize P_∞ to be ∞, we can also say:

$$(8.7)\qquad \mathcal{H}_g^{(2)} \cong \left\{\begin{array}{l}\text{set of sequences } a_1,a_2,\cdots,a_{2g+1} \\ \text{of distinct complex numbers}\end{array}\right\} \Big/ \begin{array}{l}\text{mod affine} \\ \text{equivalence} \\ a_i \longmapsto \lambda a_i+\mu\end{array}$$

hence further normalizing, e.g., P_1 to be 0, P_2 to be 1:

$$(8.8)\qquad \mathcal{H}_g^{(2)} \cong \left(\begin{array}{c}\text{open subset of } \mathbb{C}^{2g-1} \text{ of points } (a_3,a_4,\cdots,a_{2g+1}) \\ \text{such that } a_i \neq a_j,\ a_i \neq 0,1\end{array}\right)$$

This makes $\mathcal{H}_g^{(2)}$ into an affine variety. In terms of the 2nd description of $\mathcal{H}_g^{(2)}$, if t is the coordinate on $\mathbb{P}^1$, the affine ring of $\mathcal{H}_g^{(2)}$ is generated by the cross-ratios

$$\frac{t(P_i)-t(P_k)}{t(P_j)-t(P_k)} \cdot \frac{t(P_j)-t(P_\ell)}{t(P_i)-t(P_\ell)} .$$

In terms of the 3rd description (8.7) of $\mathcal{H}_g^{(2)}$, with one point normalized, the affine ring is generated by the functions

$$\frac{a_i - a_k}{a_j - a_k} .$$

Now consider the universal covering space $\hat{\mathcal{H}}_g$ of $\mathcal{H}_g^{(2)}$. Letting $\Delta_{ij} \subset \mathbb{C}^n$ be the diagonal $z_i = z_j$, and let $b \in \mathcal{H}_g^{(2)}$ be the base point $B = \{0,1,\cdots,2g,\infty\}$, we can describe it concretely as follows via (8.8):

$$(8.9) \qquad \hat{\mathcal{H}}_g \cong \left\{ \begin{array}{l} \text{space of maps } \phi: [0,1] \longrightarrow [(\mathbb{C}-(0,1))^{2g-1} - \bigcup_{i<j} \Delta_{ij}] \\ \qquad \text{such that } \phi(0) = b = (2,3,\cdots,2g) \end{array} \right\}$$

$$\begin{array}{l} \text{modulo homotopy: } \phi_0 \sim \phi_1 \text{ if} \\ \exists \Phi: [0,1]^2 \longrightarrow [(\mathbb{C}-(0,1))^{2g-1} - \bigcup \Delta_{ij}] \\ \Phi(0,t) = \phi_0(t),\ \Phi(1,t) = \phi_1(t) \\ \Phi(s,0) = b,\ \ \Phi(s,1) \text{ indep. of } s. \end{array}$$

The projection

$$\hat{\mathcal{H}}_g \longrightarrow \mathcal{H}_g^{(2)}$$

is the map $\phi \longmapsto \phi(1)$, and the covering group $\Gamma = \pi_1(\mathcal{H}_g^{(2)})$ is the group of loops in $(\mathbb{C}-(0,1))^{2g-1} - \cup\Delta_{ij}$ mod homotopy. This is essentially what E. Artin called braids with 2g+1-strands except that 2 strands are normalized at 0,1 and each strand comes back to its starting place ("pure braids"). Here is an example:

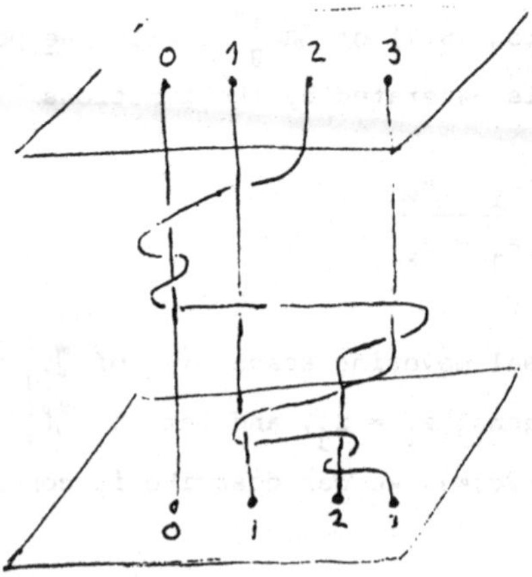

(In fact $\mathbb{Z} \times \pi_1(\mathcal{H}_g^{(2)})$ is easily shown to be $\pi_1(\mathbb{C}^{2g+1} - \cup \Delta_{ij})$, the group of all pure braids.) We call Γ the group of <u>normalized pure braids</u>. We can describe Γ a bit differently as follows:

let $G = \left\{ \begin{array}{l} \text{group of all orientation preserving} \\ \text{homeomorphisms } \phi\colon \mathbb{P}^1 \longrightarrow \mathbb{P}^1 \text{ such that} \\ \phi(0) = 0,\ \phi(1) = 1,\ \phi(\infty) = \infty, \text{ topologized in} \\ \text{the compact-open topology.} \end{array} \right\}$

let $K_g = \left\{ \begin{array}{l} \text{subgroup of } \phi \text{ such that } \phi(i) = i, \\ i = 0,1,\cdots,2g,\infty \end{array} \right\}$

Then we have a map: $\pi\colon G \longrightarrow \mathcal{H}_g^{(2)}$, $\pi\phi = \{\phi(0), \phi(1), \cdots, \phi(\infty)\}$ inducing a bijection:

(8.10) $$G/K_g \xrightarrow{\sim} \mathcal{H}_g^{(2)}$$

The following lemma is easy:

Lemma 8.11: For all $P_2,\cdots,P_{2g} \in (\mathbb{C}-(0,1))^{2g-1} - \cup\Delta_{ij}$, there are disjoint discs D_i about P_i and a map

$$\psi\colon D_2\times\cdots\times D_{2g} \longrightarrow G$$

such that $\psi(x_2,\cdots,x_{2g})(i) = x_i$. Thus ψ is a local section of π :

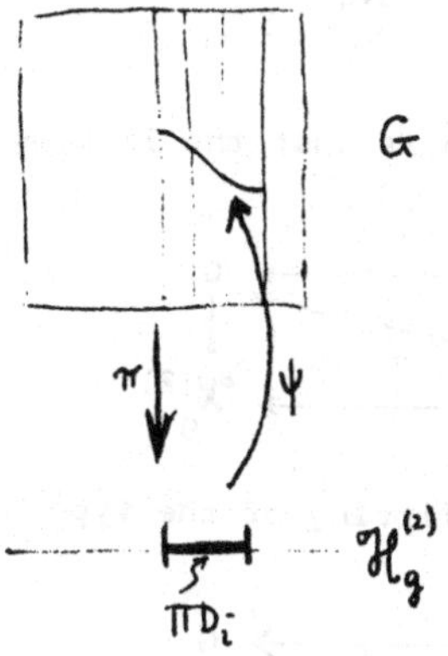

hence $\pi^{-1}(\Pi D_i) \cong \Pi D_i \times K_g$ by the group structure.

The lemma can be proven by use of suitable families of homeomorphisms of $\mathbb{P}^1$ which are different from the identity only near one point P and move P a little bit in any desired direction. The lemma implies that π has the homotopy lifting property and hence the following map p is bijective:

$$\left(\begin{array}{c}\text{space of maps}\\ \phi:[0,1]\longrightarrow G,\\ \phi(0)=\text{identity}\end{array}\right) \Bigg/ \left\{\begin{array}{l} \text{Equiv. relation } \phi_0\sim\phi_1 \text{ if}\\ \exists\Phi:\ [0,1]^2\longrightarrow G,\\ \Phi(0,t)=\phi_0(t),\Phi(1,t)=\phi_1(t)\\ \Phi(s,0)=\text{id.},\ \pi\Phi(s,1)\text{ indep. of } s\end{array}\right\}$$

$$\Big\downarrow p$$

$$\left(\begin{array}{c}\text{space of maps}\\ \phi:[0,1]\longrightarrow(\mathbb{C}-(0,1))^{2g-1}-\bigcup\Delta_{ij}\\ \phi(0)\ =\ b\end{array}\right) \Bigg/ \left\{\begin{array}{l}\text{homotopy of paths}\\ \text{with fixed initial}\\ \text{and end points}\end{array}\right\}$$

$$\cong$$

$$\hat{\mathcal{H}}_g$$

(In fact, the surjectivity of p is just the lifting of a path:

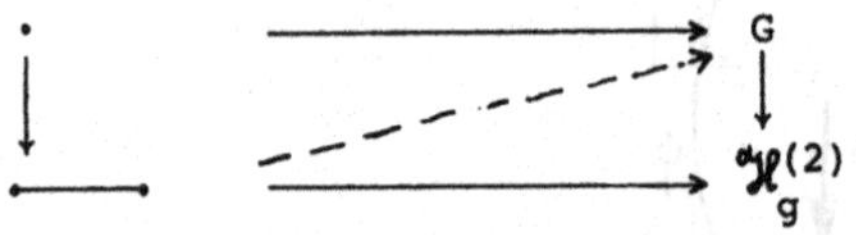

and the injectivity of p is a lifting of the type

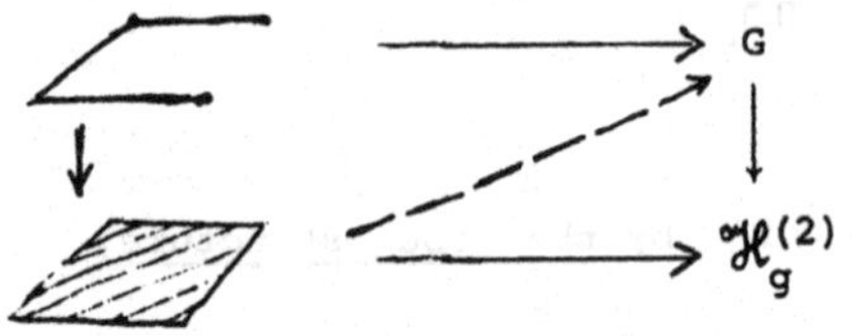

.)

Starting with $\phi: [0,1] \longrightarrow (\mathbb{C}-(0,1))^{2g-1} - \cup\Delta_{ij}$, lifting ϕ to $\psi: [0,1] \longrightarrow G$ and taking $\psi(1)$, we get a map

$$\sigma: \hat{\mathcal{H}}_g \longrightarrow G/K_g^o$$

where K_g^o is the path-connected component of the identity in K_g, i.e., $\{\phi \in K_g \mid \exists\psi: [0,1] \longrightarrow K_g \text{ such that } \psi(0) = e, \psi(1) = \phi\}$. From the homotopy lifting property of π, it follows immediately that σ is continuous. σ relates to our other constructions by a commutative diagram

$$\begin{array}{ccc} \hat{\mathcal{H}}_g & \longrightarrow & G/K_g^o \\ \downarrow & & \downarrow \\ \mathcal{H}_g^{(2)} & \xrightarrow{\approx} & G/K_g \end{array}$$

equivariant for the homomorphism:

$$\sigma_*: \Gamma \longrightarrow K_g/K_g^o$$

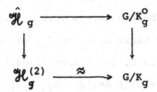

(K_g/K_g^o acts by right multiplication on G/K_g^o) given as follows: $\forall\gamma: [0,1] \longrightarrow (\mathbb{C}-(0,1))^{2g-1}-\cup\Delta_{ij}$, with $\gamma(0) = \gamma(1) = b$, lift γ via p to $\phi: [0,1] \longrightarrow G$. Then $\pi\phi(1) = b$, i.e., $\phi(1) \in K_g$. Let $\sigma_*(\gamma) = \phi(1)$. All this follows formally from the definition of σ. In fact, it can further be shown, but this is not merely formal, that $\hat{\mathcal{H}}_g$ is homeomorphic to G/K_g^o, hence $\Gamma \cong K_g/K_g^o$ but we omit this because we don't need it. The reason we have defined σ so carefully is that we wish to use σ to define the <u>global</u> period map:

$$\Omega: \hat{\mathcal{H}}_g \longrightarrow \mathfrak{h}_g.$$

In fact, given a point of $\hat{\mathcal{H}}_g$, we have 2g+2 branch points $B = \{0,1,a_2,\cdots.a_{2g},\infty\} \subset \mathbb{P}^1$ and a homeomorphism $\phi: \mathbb{P}^1 \longrightarrow \mathbb{P}^1$ such that $\phi(0) = 0$, $\phi(1) = 1$, $\phi(i) = a_i$, $2 \leq i \leq 2g$ and $\phi(\infty) = \infty$, given up to replacing ϕ by $\phi\circ\psi$, ψ fixing the i's and isotopic to the identity. Let C be the hyperelliptic curve with branch points B . Then ϕ induces a homeomorphism of the <u>standard</u> hyperelliptic curve C_0 with branch points $\{0,1,\cdots,2g,\infty\}$ with C. Taking the standard homology basis A_i,B_i on C_0, we obtain a homology basis $\phi(A_i),\phi(B_i)$ on C, hence normalized 1-forms ω_i such that $\int_{\phi(A_i)} \omega_j = \delta_{ij}$, hence finally $\Omega_{ij} = \int_{\phi(B_i)} \omega_j$. This defines the map Ω .

It should be mentioned that all the topology on the last 3 pages was traditionally compressed in the following few sentences: to each choice of branch points P, we associate a period matrix $\Omega(B)$. As B varies, we move the paths A_i,B_i continuously. $\Omega_{ij}(B)$ is locally in this way a single-valued holomorphic function on the space of B's. Globally, if we replace the space of B's by its universal covering space $\Omega_{ij}(B)$ is still a holomorphic function by analytic continuation. I'm not sure whether this "sloppy" way of talking isn't clearer!

Note that the map Ω is equivariant with respect to a homomorphism of discrete groups:

$$\Omega_*: \Gamma \longrightarrow Sp(2g,\mathbb{Z})/(\pm I) .$$

To define Ω_*, let

$$\phi: [0,1] \longrightarrow (\mathbb{C}-(0,1))^{2g-1} - \cup\Delta_{ij}$$

$$\phi(0) = \phi(1) = b$$

be a braid. Lift ϕ to

$$\Phi: [0,1] \longrightarrow G.$$

Then $\Phi(1)$ is a homeomorphism of $\mathbb{P}^1$ carrying $\{0,1,\cdots,2g,\infty\}$ to itself. Lift $\Phi(1)$ to a homeomorphism Ψ of C_0 itself. Then Ψ acts on $H_1(C,\mathbb{Z})$, in its basis $\{A_i,B_i\}$, by the $2g\times 2g$ integral symplectic matrix $\Omega_*(\phi)$. The equivariance of Ω is clear (see Ch. II, §4).

An interesting side-remark in this connection is:

Lemma 8.12. *The image of Ω_* is the level two subgroup Γ_2 of* $\gamma \in Sp(2g,\mathbb{Z})$ s.t. $\gamma \equiv I_{2g} \pmod 2$.

Proof: Note that if $\lambda_i \in H_1(C_0-B,\mathbb{Z}/2\mathbb{Z})$ is the loop around the ith branch point, then the image of λ_i in $\mathbb{P}^1-B$ goes twice around the ith branch point, hence is zero in $H_1(\mathbb{P}^1-B,\mathbb{Z}/2\mathbb{Z})$. Therefore we have a diagram:

$$\begin{array}{ccc} H_1(C_0-B,\mathbb{Z}/2\mathbb{Z})/\langle\cdots,\lambda_i,\cdots\rangle & \cong & H_1(C_0,\mathbb{Z}/2\mathbb{Z}) \\ \kappa\downarrow & & \\ H_1(\mathbb{P}^1-B,\ \mathbb{Z}/2\mathbb{Z}) & & \end{array}$$

and it's easy to check that κ is injective*. $\Phi(1)$ carries each point of B to itself, it maps a loop μ_i in $\mathbb{P}^1$ going around the ith branch point to a homologous loop. Thus $\Phi(1)$ acts by the identity on $H_1(\mathbb{P}^1-B,\ \mathbb{Z}/2\mathbb{Z})$. Thus Ψ acts by the identity on $H_1(C_0,\mathbb{Z}/2\mathbb{Z})$. Thus Im $\Omega_* \subset \Gamma_2$.

To prove the converse, recall from the Appendix to §4, Ch. II, that Γ_2, or rather its image in the group of automorphisms of $H_1(C_0,\mathbb{Z})$, is generated by the maps

$$x \longmapsto x + 2(x,e)e$$

where e is one of the elements A_i, B_i, A_i+A_j, A_i+B_j or B_i+B_j. To lift these generators in the braid group Γ, consider the following simple closed curves in $\mathbb{P}^1$-B:

---

*This is the purely topological version of the description of 2-torsion on Jac C_0 by even subsets of B. $H_1(\mathbb{P}^1-B,\mathbb{Z}/2\mathbb{Z})$ is the free group on loops μ_i around the branch points mod $\Sigma\mu_i\sim 0, 2\mu_i\sim 0$. One checks that $\kappa(A_i) = \mu_{2i-1}+\mu_{2i}$, $\kappa(B_i) = \mu_{2i}+\cdots+\mu_{2g+1}$. This proves κ injective and identifies $H_1(C_0,\mathbb{Z}/2\mathbb{Z})$ with even subsets S of B (mod S ~ B-S): let α,S correspond when $\kappa(\alpha) = \sum_{i\in S}\mu_i$.

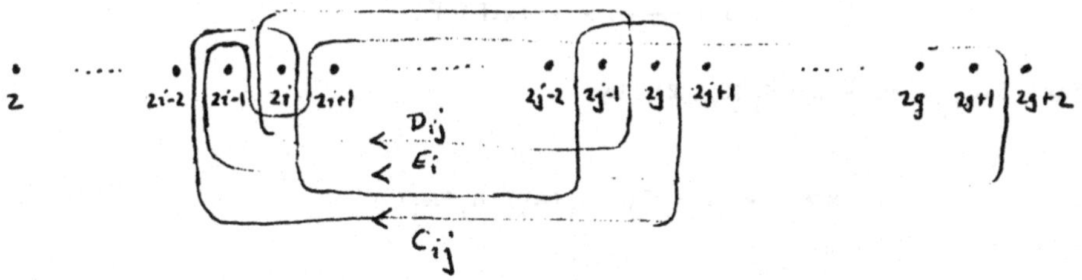

Each of these lifts in C_0 to 2 disjoint simple closed curves and a little reflection will convince the reader that they lift as follows:

i) C_{ij} lifts to $C'_{ij} \cup C''_{ij}$, $C'_{ij} \sim A_i+A_j$, $C''_{ij} \sim -(A_i+A_j)$

ii) D_{ij} lifts to $D'_{ij} \cup D''_{ij}$, $D'_{ij} \sim B_i-B_j$, $D''_{ij} \sim B_j-B_i$

iii) E_i lifts to $E'_i \cup E''_i$, $E'_i \sim A_i+B_i$, $E''_i \sim -(A_i+B_i)$.

For every simple closed curve F in $\mathbb{P}^1-B$, there is a so-called "Dehn twist" $\delta(F) \in G$: take a small collar $F \times [-\varepsilon,+\varepsilon]$ around F. Then $\delta(F)$ is the homeomorphism which is the identity outside the collar and rotates the circles $F\times\{s\}$, $-\varepsilon \le s \le \varepsilon$, through an angle $\pi(\frac{s+\varepsilon}{\varepsilon})$ varying from 0 to 2π as s varies from $-\varepsilon$ to $+\varepsilon$. Now the Dehn twist $\delta(C_{ij})$ lifts to a homeomorphism of C_0 which is

$\delta(C'_{ij})\circ\delta(C''_{ij})$. And the Dehn twist $\delta(F)$, for a path F in C_0, acts on homology by

$$x \longmapsto x + (x.F)\cdot F.$$

Thus $\delta(C_{ij})$ acts on $H_1(C_0,\mathbb{Z})$ by

$$x \longmapsto x + 2(x,A_i+A_j)(A_i+A_j).$$

Likewise:

$\delta(D_{ij})$ acts by $\quad x \longmapsto x + 2(x\cdot B_i-B_j)(B_i-B_j)$

$\delta(A_i)$ acts by $\quad x \longmapsto x + 2(x\cdot A_i)\cdot A_i$

$\delta(B_i)$ acts by $\quad x \longmapsto x + 2(x.B_i)\cdot B_i$

$\delta(E_i)$ acts by $\quad x \longmapsto x + 2(x.A_i+B_i)(A_i+B_i)$

all of which generate Γ_2. Finally, all Dehn twists $\delta(S)$ are induced by braids, i.e. $\delta(S) = \sigma_*(\gamma) \bmod K_g^0$: making S the boundary of a disc, one shrinks S to a point obtaining an isotopy of homeomorphism $\delta(S)$ with the identity. This may move 0, 1, and ∞, but by a unique projectivity, one can keep putting them back. Thus we find $\Phi: [0,1] \longrightarrow G$ with $\Phi(0) = e$, $\Phi(1) = \delta(S)$. Then $\pi\circ\Phi$ is a braid in Γ inducing $\delta(S)$. QED

Finally, having set up the spaces $\mathcal{H}_g^{(2)}, \hat{\mathcal{H}}_g$ and the map Ω, we can reformulate Thomae's theorem more geometrically. In fact, for all $\eta \in \frac{1}{2}\mathbb{Z}^{2g}$, we have the holomorphic map:

$$\begin{array}{ccc} \hat{\mathcal{H}}_g & \longrightarrow & \mathbb{C} \\ \cup\!\!\!\!\!\shortmid & & \cup\!\!\!\!\!\shortmid \\ P & \longmapsto & \vartheta[\eta](0,\Omega(P)) \end{array}$$

Either by the functional equation for $\vartheta(z,\Omega)$, or by Thomae's formula, we see that the functions

$$\frac{\vartheta[\eta](0,\Omega)^4}{\vartheta[0](0,\Omega)^4}$$

on $\hat{\mathcal{H}}_g$ are Γ-invariant, hence are holomorphic functions on $\mathcal{H}_g^{(2)}$, depending only on $\delta \in \frac{1}{2}\mathbb{Z}^{2g}/\mathbb{Z}^{2g}$. Thomae's formula implies:

<u>Corollary 8.13</u>: <u>The affine ring of $\mathcal{H}_g^{(2)}$ is generated by the nowhere zero functions</u>:

$$\left(\frac{\vartheta[\eta_S](0,\Omega)}{\vartheta[0]\ (0,\Omega)}\right)^{\pm 4}, \quad S \subset B \quad \underline{\text{such that}}\ \#U\circ S = g+1.$$

<u>Proof:</u> Normalizing one branch point to ∞, and letting $a_1,a_2,\cdots,a_{2g+1}$ be the others, we must check that each atio a_k-a_ℓ/a_k-a_m is a polynomial in these 4th powers. We use the identity:

$$\left(\frac{a_k-a_\ell}{a_k-a_m}\right)^2 - \left(\frac{a_\ell-a_m}{a_k-a_m}\right)^2 + 1 = 2\left(\frac{a_k-a_\ell}{a_k-a_m}\right).$$

If we write $\{1,2,\cdots,2g+1\} = V_1 \amalg V_2 \amalg \{k\}$, $\#V_1 = \#V_2 = g$, then by

Thomae's formula

$$\frac{\prod_{i\in V_1}(a_k-a_i)}{\prod_{i\in V_2}(a_k-a_i)} = \pm \frac{\vartheta[\eta_{(V_2+k)\circ U}](0,\Omega)^4}{\vartheta[\eta_{(V_1+k)\circ U}](0,\Omega)^4} \tag{8.14}$$

Write instead $\{1,2,\cdots,2g+1\} = V_3 \amalg V_4 \amalg \{k,\ell,m\}$, $\#V_3 = \#V_4 = g-1$ and apply (8.14) to the pairs $V_1 = V_3+\ell$, $V_2 = V_4+m$ and then to $V_1 = V_3+m$, $V_2 = V_4+\ell$. Dividing we find

$$\left(\frac{a_k-a_\ell}{a_k-a_m}\right)^2 = \pm \frac{\vartheta[\eta_1]^4\,\vartheta[\eta_2]^4}{\vartheta[\eta_3]^4\,\vartheta[\eta_4]^4}$$

for suitable η_i. <u>QED</u>.

The relations among these generators presumably may all be derived from various specializations of Frobenius' identity.

§9. Characterization of hyperelliptic period matrices

The goal of this section is to prove that the fundamental Vanishing property of §6 characterizes hyperelliptic Jacobians. The method will be to show that any abelian variety X_Ω which has the Vanishing property must have a covering of degree 2^{g+1} which occurs as an orbit of the g commuting flows of the Neumann dynamical system.

To state the result precisely, we fix, as above, the following notation:

B = fixed set with 2g+2 elements in it

$U \subset B$, a subset of g+1 elements in it.

$$\eta: \left\{ \begin{matrix} \text{group of subsets} \\ T \subset B, \#T \text{ even} \\ \text{mod } T \sim CT \end{matrix} \right\} \xrightarrow{\approx} \tfrac{1}{2}\mathbb{Z}^{2g}/\mathbb{Z}^{2g}$$

where η is any isomorphism satisfying

$$\#(T_1 \cap T_2) \equiv \left[{}^t(2\eta'(T_1)) \cdot (2\eta''(T_2)) - {}^t(2\eta''(T_1)) \cdot (2\eta'(T_2)) \right] \bmod 2$$

$$\frac{\#(T \circ U) - (g+1)}{2} \equiv {}^t(2\eta'(T)) \cdot (2\eta''(T)) \bmod 2$$

where $\eta(T) = (\eta'(T), \eta''(T))$. We shall subsequently abbreviate $\eta(T)$ by η_T.

<u>Theorem 9.1</u>: <u>Assume</u> $\Omega \in \mathfrak{H}_g$ <u>satisfies</u>

$$\vartheta[\eta_T](0,\Omega) = 0 \iff \#(T \circ U) \neq g+1 .$$

<u>Then</u> Ω <u>is the period matrix of a smooth hyperelliptic curve of genus</u> g.

<u>Proof</u>: First of all, to write our formulae with unambiguous signs, we are forced to make a choice of lifting of η_T from $\frac{1}{2}\mathbb{Z}^{2g}/\mathbb{Z}^{2g}$ to $\frac{1}{2}\mathbb{Z}^{2g}$. To do this, we choose a fixed element $\infty \in B-U$, and choose

$$\eta_i \in \frac{1}{2}\mathbb{Z}^{2g}, \qquad \eta_i \equiv \eta_{\{i,\infty\}} \bmod \mathbb{Z}^{2g}$$

for all $i \in B-\infty$. We also set $\eta_\infty = 0$. Then for all T define a lifting $\eta_T \in \frac{1}{2}\mathbb{Z}^{2g}$ by

$$\eta_T = \sum_{i \in T} \eta_i .$$

(The "standard" choice, if $B = \{1,\cdots,2g+1,\infty\}$, is

$$\eta_{2i-1} = \begin{pmatrix} 0\cdots 0 \ \frac{1}{2} \ 0\cdots 0 \\ \frac{1}{2}\cdots\frac{1}{2} \ 0 \ 0\cdots 0 \end{pmatrix}$$

$\updownarrow$ $i^{\underline{th}}$ place

$$\eta_{2i} = \begin{pmatrix} 0\cdots 0 \ \frac{1}{2} \ 0\cdots 0 \\ \frac{1}{2}\cdots\frac{1}{2} \ \frac{1}{2} \ 0\cdots 0 \end{pmatrix}$$

but there is no need to get <u>that</u> specific.)

The first part of the proof is to investigate the differential of the theta function $\vartheta[\eta_T](z,\Omega)$ at $z = 0$. The tool at our disposal is Frobenius' formula, and we propose to differentiate it, and substitute so that very few terms remain. In formula

(F_{ch}) in theorem 7.1, replace x_1 by x_1+y, x_2 by x_2-y, take the differential with respect to y and set y = 0. We get, assuming $\sum a_i = \sum z_i = 0$:

$$(F^2_{ch}) \sum_{j\in B} \varepsilon_U(j) \Big(d\vartheta[a_1+\eta_j](z_1) \cdot \vartheta[a_2+\eta_j](z_2) -$$

$$d\vartheta[a_2+\eta_j](z_2) \cdot \vartheta[a_1+\eta_j](z_1) \Big) \cdot \vartheta[a_3+\eta_j](z_3) \cdot \vartheta[a_4+\eta_j](z_4) = 0.$$

Note first that $\vartheta[\eta_S](z,\Omega)$ is an even function if

$$\#(S\circ U) \equiv g+1 \bmod 4$$

and is odd if

$$\#(S\circ U) \equiv g-1 \bmod 4.$$

Therefore

$$d\vartheta[\eta_S](0,\Omega) = 0$$

if $\#(S\circ U) \equiv g+1 \bmod 4$ and we may restrict our attention to the case $\#(S\circ U) \equiv g-1 \bmod 4$, and, replacing S by CS if necessary, $\#(S\circ U) \leq g-1$.

<u>Lemma 9.2</u>: $d\vartheta[\eta_S](0,\Omega) = 0$ <u>if</u> $\#(S\circ U) \equiv (g-1) \bmod 4$ <u>and</u> $\#(S\circ U) < (g-1)$.

<u>Proof</u>: Let $T \subset B$ satisfy $\#T = g-5$. In (F^2_{ch}), let $z_i = 0$, all i. Moreover, take $A,B,C \subset B-T$ 3 disjoint sets of 3 elements each, let $a \in A$ be one of its elements and set

$$a_1 = \eta_{T\circ U} + \eta_a$$

$$a_2 = \eta_{(T+A+B)\circ U} + \eta_a$$

$$a_3 = \eta_{(T+A+C)\circ U} + \eta_a$$

$$a_4 = -a_1-a_2-a_3$$

$$\equiv \eta_{(T+B+C)\circ U} + \eta_a \bmod \mathbb{Z}^{2g} .$$

Then all terms with a factor $\vartheta[a_1+\eta_i](0)$ are zero, so the $d\vartheta$ goes with the $1^{\underline{st}}$ factor. For the last 3 factors to be non-zero, we need

$$\#(T+A+C-a)\circ\{i\} = g+1,$$

$$\#(T+A+C-a)\circ\{i\} = g+1,$$

$$\text{and} \quad \#(T+B+C+a)\circ\{i\} = g+1.$$

This only happens if $i = a$. So the formula reduces to

$$d\vartheta[\eta_{T\circ U}](0)\cdot\vartheta[\eta_{(T+A+B)\circ U}](0)\cdot\vartheta[\eta_{(T+A+C)\circ U}]\cdot\vartheta[\eta_{(T+B+C)\circ U}](0) = 0.$$

Since the last three are non-zero, the $1^{\underline{st}}$ is zero. A similar argument shows that $d\vartheta[\eta_{T\circ U}](0) = 0$ if $T \subset B$ satisfies $\#T = g-9, g-13$, etc. QED

Lemma 9.3. For all $R \subset B$, $\#R = g-2$, and elements $a,b,c \in B-R$, there is a relation

$$\lambda\cdot d\vartheta[\eta_{(R+a)\circ U}](0) + \mu\cdot d\vartheta[\eta_{(R+b)\circ U}](0) + \nu\cdot d\vartheta[\eta_{(R+c)\circ U}](0) = 0$$

where $\lambda\mu\nu \neq 0$.

<u>Proof</u>: Let d,e be 2 elements of B-R-{a,b,c} and let $f \in R$. In (F^2_{ch}), let $z_i = 0$, all i, and let

$$a_1 = \eta_{(R-f)\circ U} + \eta_f$$

$$a_2 = \eta_{(R-f+d+e)\circ U} + \eta_f$$

$$a_3 = \eta_{(R-f+a+b+c+d)\circ U} + \eta_f$$

$$a_4 = -a_1-a_2-a_3$$

$$\equiv \eta_{(R-f+a+b+c+e)\circ U} + \eta_f .$$

Then all terms with a factor $\vartheta[a_1+\eta_i](0)$ are zero, so in each term the $d\vartheta$ goes with the first factor. For the last 3 factors to be non-zero, we need

$$\#(R+d+e)\circ\{i\} = g+1$$
$$\#(R+a+b+c+d)\circ\{i\} = g+1$$
$$\#(R+a+b+c+e)\circ\{i\} = g+1.$$

This happens if i = a, b, or c, giving 3 remaining terms and a formula just as required. <u>QED</u>

<u>Lemma 9.4</u>: <u>Let</u> $S,T \subset B$, $\#S = g$, $\#T = g-1$. <u>Then in</u> $T^*_{X_\Omega,0}$,

$$d\vartheta[\eta_{T\circ U}](0) \in \left\{\begin{array}{c}\text{span of differentials } d\vartheta[\eta_{(S-i)\circ U}](0),\\ \text{for all } i\in S-S\cap T\end{array}\right\}$$

Proof: Prove this by induction on $\#(S\cap T)$. If, to start with, $S = R+a+b$, $T = R+c$, ($a,b,c \in B-R$ distinct, and $\#R = g-2$), then the result is precisely lemma 9.3. In general, choose $a,b \in S-S\cap T$ and $c \in T-S\cap T$. By lemma 9.3,

$$d\vartheta[\eta_{T\circ U}](0) = \lambda\cdot d\vartheta[\eta_{(T-c+a)\circ U}](0) + \mu\cdot d\vartheta[\eta_{(T-c+b)\circ U}](0).$$

Now $(T-c+a)$ and $(T-c+b)$ both have one more element in common with S than T did, so by induction $d\vartheta[\eta_{(T-c+a)\circ U}](0)$ and $d\vartheta[\eta_{(T-c+b)\circ U}](0)$ are both in the required span. Therefore, so is $d\vartheta[\eta_{T\circ U}](0)$. QED

Lemma 9.5: Let $S \subset B$, $\#S = g$. Then the g differentials $\omega_a = d\vartheta[\eta_{(S-a)\circ U}](0)$, $a \in S$, span $T^*_{X_\Omega,0}$.

Proof: We use the fact that the abelian variety X_Ω is embedded in projective space by the functions $\vartheta[\eta](2z,\Omega)$, when η runs over coset representatives of $\frac{1}{2}\mathbb{Z}^{2g}/\mathbb{Z}^{2g}$ (see Ch. 2, §). It follows that the whole set of differentials $d\vartheta[\eta_S](0,\Omega)$ must span $T^*_{X_\Omega,0}$, as S runs over all even subsets of B. By lemma 9.2 we may as well assume $\#(S\circ U) = g-1$. By lemma 9.4, the whole space is still spanned by $d\vartheta[\eta_S](0,\Omega)$'s where $S = (S_0-\{a\})\circ U$, S_0 is any one set of g elements of B and a runs over the elements of S_0. QED

Lemma 9.6. For all $a \in B$, there is a unique vector $D_a \in T_{X_\Omega,0}$, up to a scalar, such that for all $T \subset B$, $\#T = g-1$

$$D_a\vartheta[\eta_{T\circ U}](0) = 0 \iff a \in T.$$

<u>Proof</u>: Fix a subset $S \subset B$ with $\#S = g$, $a \in S$. Then the requirement

$$D_a \vartheta[\eta_{(S-b)\circ U}](0) = 0, \quad \text{all} \quad b \in S-\{a\}$$

determines D_a up to a scalar by lemma 9.5. To see that $D_a\vartheta[\eta_{T\circ U}](0) = 0$ if $a \in T$, $\#T = g-1$, use lemma 9.4. On the other hand, if $D_a\vartheta[\eta_{T\circ U}](0) = 0$ when $a \notin T$, $\#T = g-1$, we would have that the differentials $d\vartheta[\eta_{T+a-b}](0)$, all $b \in T+a$, were linearly dependent, contradicting lemma 9.5.

We now concentrate on the vectors D_k, $k \in U$, and D_∞. By 9.5, no g of the vectors $\{D_k | k \in U\}$ lie in a hyperplane, so we may normalize the whole set up to multiplication of the whole set by a single scalar by requiring

$$\sum_{k \in U} D_k = 0.$$

Then define scalars a_k, $k \in U$, by:

$$D_\infty = \sum a_k D_k.$$

Note that for fixed D_∞, $\{D_k\}$, the a_k are determined up to a substitution $a_k \longmapsto a_k+\mu$; and if D_∞, $\{D_k\}$ are changed by scalars, the a_k change by an affine substitution $a_k \longmapsto \lambda a_k+\mu$. So far the proof is quite natural. We must, however, normalize $D_\infty, \{D_k\}$ a bit more and for this a rather <u>ad hoc</u> Corollary of Frobenius's formula is needed.

<u>Lemma 9.7</u>: <u>For all</u> $j,k,\ell \in U$ <u>distinct</u>,

$$e^{4\pi i\eta_\ell'\cdot\eta_j''}\,\vartheta[\eta_j](0)^2\cdot D_\infty\vartheta[\eta_{\{\ell,k\}}](0)\cdot D_\ell\vartheta[\eta_{\{\ell,k\}}](0) =$$

$$e^{4\pi i\,{}^t\eta_\ell'\cdot\eta_k''}\cdot\vartheta[\eta_k](0)^2\cdot D_\infty\vartheta[\eta_{\{\ell,j\}}](0)\cdot D_\ell\vartheta[\eta_{\{\ell,j\}}](0) .$$

Proof: Start with the formula

$$\sum_{i\in U+\infty} \varepsilon_\infty(i)\vartheta[\eta_i](0)^2\cdot\vartheta[\eta_i](z)^2 = 0$$

(Corollary 7.4). Replace z by $z + \Omega\eta'_{\{j,k,\ell,\infty\}} + \eta''_{\{j,k,\ell,\infty\}}$ and it becomes:

$$\sum_{i\in U+\infty} \varepsilon_\infty(i)e^{-4\pi i\,{}^t\eta'_{\{j,k,\ell,\infty\}}\cdot\eta''_i}\vartheta[\eta_i](0)^2\cdot\vartheta[\eta_i+\eta_{\{j,k,\ell,\infty\}}](z)^2 = 0.$$

Differentiate this first by D_∞, second by D_ℓ and set $z = 0$:

$$\sum_{i\in U+\infty} \varepsilon_\infty(i)e^{-4\pi i\,{}^t\eta'_{\{j,k,\ell,\infty\}}\cdot\eta''_i}\vartheta[\eta_i](0)^2\cdot\Big[D_\infty\vartheta[\eta_i+\eta_{\{j,k,\ell,\infty\}}](0)\cdot D_\ell\vartheta[\eta_i+\eta_{\{j,k,\ell,\infty\}}](0)$$

$$+\vartheta[\eta_i+\eta_{\{j,k,\ell,\infty\}}](0)\cdot D_\infty D_\ell\vartheta[\eta_i+\eta_{\{j,k,\ell,\infty\}}](0)\Big] = 0\,.$$

Since the sets $U\circ\{j,k,\ell\}\circ\{i\}$ all have at most g-1 elements, $\vartheta[\eta_i+\eta_{\{j,k,\ell,\infty\}}](0) = 0$ for all $i\in B$. Moreover, if $i\in U+\infty$, $\#(U\circ\{j,k,\ell\}\circ\{i\}) = g-1$ only if $i = j,k,\ell$ or ∞. To get a non-zero term in the above formula, we also need

$$\infty,\ell \notin U\circ\{j,k,\ell\}\circ\{i\}$$

which narrows down the possibilities to i = j or k. We get

$$e^{4\pi i\,{}^t\eta'_{\{j,k,\ell,\infty\}}\cdot\eta''_k}\vartheta[\eta_k](0)^2\cdot D_\infty\vartheta[\eta_{\{j,\ell\}}](0)\cdot D_\ell\vartheta[\eta_{\{j,\ell\}}](0)$$

$$+\,e^{4\pi i\,{}^t\eta'_{\{j,k,\ell,\infty\}}\cdot\eta''_j}\vartheta[\eta_j](0)^2\cdot D_\infty\vartheta[\eta_{\{k,\ell\}}](0)\cdot D_\ell\vartheta[\eta_{\{k,\ell\}}](0) = 0\,.$$

Using the fact that

$$-1 = (-1)^{\#\{j,\infty\}\cap\{k,\infty\}} = e^{4\pi i({}^t\eta_j'\eta_k'' - {}^t\eta_k'\cdot\eta_j'')}$$

and

$$+1 = (-1)^{\frac{\#(\{j,\infty\}\circ U)-g-1}{2}} = e^{4\pi i\,{}^t\eta_j'\cdot\eta_j''}$$

$$+1 = (-1)^{\frac{\#(\{k,\infty\}\circ U)-g-1}{2}} = e^{4\pi i\,{}^t\eta_k'\cdot\eta_k''},$$

the two signs may be replaced by $e^{4\pi i\,{}^t\eta_\ell'\cdot\eta_k''}$ and $e^{4\pi i\,{}^t\eta_\ell'\cdot\eta_j''}$, respectively.

QED

Corollary 9.8. Replacing D_∞ by $\lambda\cdot D_\infty$, $\lambda\in\mathbb{C}^*$, we can assume

$$\vartheta[0](0)^2\cdot D_\infty\vartheta[\eta_{\{\ell,k\}}](0)\cdot D_\ell\vartheta[\eta_{\{\ell,k\}}](0) = e^{4\pi i\,{}^t\eta_k'\cdot\eta_\ell''}\cdot\vartheta[\eta_\ell](0)^2\cdot\vartheta[\eta_k](0)^2$$

for all $\ell,k\in U$.

Proof: If we choose D_∞ suitably, this will be true for one pair ℓ_0,k_0. Now vary k. By lemma 9.7, the formula is true for ℓ_0 and all k. Now interchange ℓ_0 and k. Since $\sum_{j\in U} D_j = 0$, we get

$$0 = \sum_{j\in U} D_j\vartheta[\eta_{\{\ell,k\}}](0) = D_k\vartheta[\eta_{\{\ell,k\}}](0) + D_\ell\vartheta[\eta_{\{\ell,k\}}](0).$$

Since

$$\frac{e^{4\pi i\,{}^t\eta_\ell'\cdot\eta_k''}}{e^{4\pi i\,{}^t\eta_k'\,\eta_\ell''}} = (-1)^{\#(\{\ell,\infty\}\cap\{k,\infty\})} = -1,$$

the formula is also true for k and ℓ_0. Now varying ℓ_0, it is always true. QED

The next step in the proof is an elegant and quite important consequence of Frobenius's formula:

<u>Proposition 9.9.</u> <u>Let</u> $T \subset B$ <u>satisfy</u> $\#T = g-1$ <u>and let</u> a,b,c <u>be</u> 3 <u>distinct elements in</u> $B-T$. <u>Then</u>

$$\vartheta[\eta_{T\circ U}+\eta_{\{a,c\}}](0)\cdot\vartheta[\eta_{T\circ U}+\eta_{\{b,c\}}](0)\cdot\left[D_c\vartheta[\eta_a](z)\cdot\vartheta[\eta_b](z)-D_c\vartheta[\eta_b](z)\cdot\vartheta[\eta_a](z)\right]$$

$$= \sigma\cdot D_c\vartheta[\eta_{T\circ U}](0)\cdot\vartheta[\eta_{T\circ U}+\eta_{\{a,b\}}](0)\cdot\vartheta[\eta_c](z)\cdot\vartheta[\eta_{\{a,b,c,\infty\}}](z)\ ,$$

where $\sigma = e^{4\pi i\,{}^t\eta'_a\cdot\eta''_b}\cdot e^{4\pi i\eta'_{T\circ U}\cdot\eta''_c} = \pm 1$.

<u>Proof</u>: In formula (F^2_{ch}), set $z_1 = z_4 = 0$, $z_2 = z$, $z_3 = -z$. Moreover, set

$$a_1 = \eta_{T\circ U} + \eta_c$$

$$a_2 = 0$$

$$a_3 = \eta_{\{a,b\}}$$

$$a_4 = -a_1-a_2-a_3 \equiv \eta_{T\circ U}+\eta_{\{a,b,c,\infty\}} \bmod \mathbb{Z}^{2g}\ .$$

Finally, evaluate the differential on the vector D_c.

The coefficients in the $j^{\underline{th}}$ term of (F^2_{ch}) are, up to constants

$$D_c\vartheta[\eta_{T\circ U}+\eta_c+\eta_j](0)\cdot\vartheta[\eta_{T\circ U}+\eta_{\{a,b,c,\infty\}}+\eta_j](0)$$

and

$$\vartheta[\eta_{T\circ U}+\eta_c+\eta_j](0)\cdot\vartheta[\eta_{T\circ U}+\eta_{\{a,b,c,\infty\}}+\eta_j](0).$$

For the $1^{\underline{st}}$ to be non-zero we need

$$\#(T+c)\circ\{j\} = g-1, \quad c \notin (T+c)\circ\{j\}$$

by lemma 9.6. This means $j = c$. For the $2^{\underline{nd}}$ to be non-zero, we need

$$\#(T+c)\circ\{j\} = g+1$$

$$\#(T+a+b+c)\circ\{j\} = g+1$$

which means $j = b$ or c. Writing out the three non-zero terms and evaluating the sign with some pain, we get the result. QED

We are now ready for the key point of the proof. We define a 2^{g+1}-sheeted covering X'_Ω of X_Ω and a morphism

$$\phi: X'_\Omega - V(\vartheta[0]) \longrightarrow \mathbb{C}^{2g+2}$$

as follows:

$$X'_\Omega = \mathbb{C}^g/L'_\Omega$$

$$L'_\Omega = \{\Omega p+q \mid p,q \in \mathbb{Z}^g \text{ and } {}^t\eta'_i q - {}^t\eta''_i p \in \mathbb{Z}, \text{ all } i \in U\}$$

If $\{x_i,\ i \in U;\ y_i,\ i \in U\}$ are coordinates on $\mathbb{C}^{2g+2}$

ϕ is defined by

$$x_i = \frac{\vartheta[\eta_i](0)\cdot\vartheta[\eta_i](z)}{\vartheta[0](0)\cdot\vartheta[0](z)}, \quad i \in U$$

$$y_i = D_\infty\left(\frac{\vartheta[\eta_i](0)\cdot\vartheta[\eta_i](z)}{\vartheta[0](0)\cdot\vartheta[0](z)}\right), \quad i \in U.$$

Note that $2L_\Omega \subset L'_\Omega \subset L_\Omega$ and $[L_\Omega : L'_\Omega] = 2^{g+1}$, and that L'_Ω is precisely the lattice with respect to which all the functions

$\vartheta[\eta_i](z)/\vartheta[0](z)$, $i \in U$, are periodic. Moreover, by Corollary 7.5, the image of ϕ lies in the affine variety

$$\sum x_i^2 = 1 ,$$

hence by differentiating, the image lies in

$$(\sum x_i^2) - 1 = \sum x_i y_i = 0$$

– called $T_{\mathbb{C}}(S^g)$ in §4, the complexified tangent bundle to S^g. What we shall prove now is that the vector fields D_k, $k \in U$, on X_Ω' are mapped to half the Hamiltonian vector fields X_{F_k} on $\mathbb{C}^{2g+2}$ defined in §4. It will follow that ϕ is an isogeny of the torus X_Ω' onto one of the tori obtained by simultaneously integrating the X_{F_k}, which by the theory of §4 are precisely 2^{g+1}-fold covers of the hyperelliptic jacobians. It will then follow easily that, in fact, X_Ω is isomorphic to the corresponding jacobian.

Recall that $F_k(X,Y)$, $k \in U$, are the functions:

$$F_k(x,y) = x_k^2 + \sum_{\substack{\ell \neq k \\ \ell \in U}} \frac{(x_k y_\ell - x_\ell y_k)^2}{a_k - a_\ell}$$

(the a_k here are the same a_k defined earlier in this proof). The corresponding vector fields X_{F_k} are given by:

$$X_{F_k}(x_\ell) = \frac{2(x_k y_\ell - x_\ell y_k)}{a_k - a_\ell} \cdot x_k, \qquad \text{if } \ell \neq k$$

$$= \sum_{\substack{p \neq k \\ p \in U}} \frac{2(x_k y_p - x_p y_k)}{a_k - a_p} (-x_p) \qquad \text{if } \ell = k$$

and

$$X_{F_k}(y_\ell) = \frac{2(x_k y_\ell - x_\ell y_k)}{a_k - a_\ell} \cdot y_k + 2x_\ell x_k^2 \qquad \text{if } \ell \neq k$$

$$= \sum_{\substack{p \neq k \\ p \in U}} \frac{2(x_k y_p - x_p y_k)}{a_k - a_p} (-y_p) + 2x_k(x_k^2 - 1) \qquad \text{if } \ell = k$$

(See §4, Proof of Theorem 4.7.)

Note that $\sum X_{F_k} = X_{\Sigma F_k} = X_1 = 0$. Now let capital X_i be the function on X'_Ω:

$$X_i = \frac{\vartheta[\eta_i](0) \cdot \vartheta[\eta_i](z)}{\vartheta[0](0) \cdot \vartheta[0](z)}$$

and let capital Y_i by $D_\infty X_i$, again a function on X'_Ω. What we claim is that if we substitute D_k for X_{F_k}, X_k for x_k, Y_k for y_k, then if $\ell \neq k$, $D_k(X_\ell)$, $D_k(Y_\ell)$ are given by almost the same formulae on X'_Ω:

<u>Lemma</u> 9.10. <u>If $\ell \neq k$, then on X'_Ω:</u>

$$D_k(X_\ell) = \frac{(X_k Y_\ell - X_\ell Y_k)}{a_k - a_\ell} \cdot X_k$$

$$D_k(Y_\ell) = \frac{(X_k Y_\ell - X_\ell Y_k)}{a_k - a_\ell} \cdot Y_k + X_\ell X_k^2 \quad .$$

But $\sum D_k = 0$, so $D_k(X_k), D_k(Y_k)$ are also given by the same formulae as $\frac{1}{2} X_{F_k}(x_k)$, $\frac{1}{2} X_{F_k}(y_k)$. Hence the lemma implies:

<u>Corollary</u> 9.11: <u>The differential of ϕ carries the vector field D_k to the vector field $\frac{1}{2} X_{F_k}$,</u>

Before proving lemma 9.10, we shall evaluate $X_k Y_\ell - X_\ell Y_k$ in simpler terms:

<u>Lemma</u> 9.11. If $\ell \neq k$,

$$X_k Y_\ell - X_\ell Y_k = e^{4\pi i \, {}^t\eta'_k \cdot \eta''_\ell} \cdot \frac{D_\infty \vartheta[\eta_{\{k,\ell\}}](0)}{\vartheta[0](0)} \cdot \frac{\vartheta[\eta_{\{k,\ell\}}](z)}{\vartheta[0](z)} \quad .$$

<u>Proof:</u> $$X_k Y_\ell - X_\ell Y_k = X_k^2 D_\infty (X_\ell / X_k)$$

$$= \frac{\vartheta[\eta_k](0)\vartheta[\eta_\ell](0)}{\vartheta[0](0)^2} \cdot \frac{\vartheta[\eta_k](z) \cdot D_\infty \vartheta[\eta_\ell](z) - \vartheta[\eta_\ell](z) \cdot D_\infty \vartheta[\eta_k](z)}{\vartheta[(0)](z)^2} \quad .$$

Using Proposition 9.9, with $T = U - \{k, \ell\}$, $c = \infty$, $a = \ell$, $b = k$, the second term on the right equals

$$e^{4\pi i\,{}^t\eta'_\ell\cdot\eta''_k}\cdot\frac{D_\infty\vartheta[\eta_{\{k,\ell\}}](0)\cdot\vartheta[2\eta_{\{k,\ell\}}](0)}{\vartheta[\eta_k+2\eta_\ell](0)\cdot\vartheta[\eta_\ell+2\eta_k](0)}\cdot\frac{\vartheta[\eta_{\{k,\ell\}}](z)}{\vartheta[0](z)}.$$

Simplifying the characteristics and working out the sign, this gives Lemma 9.11. QED

Proof of 9.10: If $\ell \neq k$,

$$D_k(X_\ell) = \frac{\vartheta[\eta_\ell](0)}{\vartheta[0](0)}\cdot\frac{\vartheta[0](z)\cdot D_k\vartheta[\eta_\ell](z)-\vartheta[\eta_\ell](z)\cdot D_k\vartheta[0](z)}{\vartheta[0](z)^2}.$$

Using Proposition 9.9 with $T = U-\{k,\ell\}$, $c = k$, $a = \ell$, $b = \infty$, the second term on the right equals

$$e^{4\pi i\,{}^t\eta'_{\{k,\ell\}}\cdot\eta''_k}\cdot\frac{D_k\vartheta[\eta_{\{k,\ell\}}](0)\cdot\vartheta[\eta_k+2\eta_\ell](0)}{\vartheta[2\eta_k+2\eta_\ell](0)\cdot\vartheta[2\eta_k+\eta_\ell](0)}\cdot\frac{\vartheta[\eta_k](z)\vartheta[\eta_{\{k,\ell\}}](z)}{\vartheta[0](z)^2},$$

hence

$$D_k(X_\ell) = e^{4\pi i\,{}^t\eta'_k\eta''_\ell}\,\frac{D_k\vartheta[\eta_{\{k,\ell\}}](0)\cdot\vartheta[\eta_k](0)}{\vartheta[0](0)^2}\cdot\frac{\vartheta[\eta_k](z)\cdot\vartheta[\eta_{\{k,\ell\}}](z)}{\vartheta[0](z)^2}.$$

On the other hand,

$$D_\infty = \sum_{p\in U} a_pD_p = \sum_{p\in U}(a_p-a_\ell)D_p,$$

hence

$$D_\infty\vartheta[\eta_{\{k,\ell\}}](0) = \sum_{p\in U}(a_p-a_\ell)D_p\vartheta[\eta_{\{k,\ell\}}](0).$$

But $D_p\vartheta[\eta_{\{k,\ell\}}](0) = 0$ if $p \neq k,\ell$ (because $p \in U\circ\{k,\ell\}$), so

$$D_\infty\vartheta[\eta_{\{k,\ell\}}](0) = (a_k - a_\ell) D_k\vartheta[\eta_{\{k,\ell\}}](0).$$

Therefore,

$$D_k(X_\ell) = \frac{1}{a_k - a_\ell} \cdot \left(e^{4\pi i\, {}^t\eta_k' \cdot \eta_\ell''} \cdot \frac{D_\infty\vartheta[\eta_{\{k,\ell\}}](0) \cdot \vartheta[\eta_{\{k,\ell\}}](z)}{\vartheta[0](0) \cdot \vartheta[0](z)}\right) \cdot \left(\frac{\vartheta[\eta_k](0)\,\vartheta[\eta_k](z)}{\vartheta[0](0) \cdot \vartheta[0](z)}\right)$$

$$= \frac{X_k Y_\ell - X_\ell Y_k}{a_k - a_\ell} \cdot X_k \qquad \text{by Lemma 9.11.}$$

Finally,

$$\begin{aligned} D_k(Y_\ell) &= D_\infty(D_k(X_\ell)) \\ &= D_\infty\left(X_k \cdot \frac{X_k Y_\ell - X_\ell Y_k}{a_k - a_\ell}\right) \\ &= Y_k \cdot \frac{X_k Y_\ell - X_\ell Y_k}{a_k - a_\ell} + \frac{X_k}{a_k - a_\ell} \cdot D_\infty(X_k Y_\ell - X_\ell Y_k); \end{aligned}$$

hence it remains to prove

$$D_\infty(X_k Y_\ell - X_\ell Y_k) = (a_k - a_\ell) X_\ell X_k.$$

But

$$D_\infty(X_k Y_\ell - X_\ell Y_k) = e^{4\pi i\, {}^t\eta_k' \cdot \eta_\ell''} \frac{D_\infty\vartheta[\eta_{\{k,\ell\}}](0)}{\vartheta[0](0)} \cdot \frac{\vartheta[0](z) D_\infty\vartheta[\eta_{\{k,\ell\}}](z) - \vartheta[\eta_{\{k,\ell\}}](z) \cdot D_\infty\vartheta[0](z)}{\vartheta[0](z)^2}$$

Now in the proof of 9.11, we deduced as a special case of 9.9 that

$$\vartheta[\eta_k](z)\cdot D_\infty\vartheta[\eta_\ell](z) - \vartheta[\eta_\ell](z)\cdot D_\infty\vartheta[\eta_k](z)$$

$$= e^{4\pi i\,{}^t\eta_k'\,\eta_\ell''}\,\frac{D_\infty\vartheta[\eta_{\{k,\ell\}}](0)\cdot\vartheta[0](0)}{\vartheta[\eta_k](0)\cdot\vartheta[\eta_\ell](0)}\cdot\vartheta[\eta_{\{k,\ell\}}](z)\cdot\vartheta[0](z).$$

Substituting $z+\Omega\eta_k'+\eta_k''$ for z and rewriting the theta functions, this gives

$$\vartheta[0](z)\cdot D_\infty\vartheta[\eta_{\{k,\ell\}}](z) - \vartheta[\eta_{\{k,\ell\}}](z)\cdot D_\infty\vartheta[0](z)$$

$$= -\frac{D_\infty\vartheta[\eta_{\{k,\ell\}}](0)\,\vartheta[0](0)}{\vartheta[\eta_k](0)\,\vartheta[\eta_\ell](0)}\cdot\vartheta[\eta_\ell](z)\cdot\vartheta[\eta_k](z)\,.$$

(The minus sign comes from

$$e^{4\pi i\,{}^t\eta_k'\cdot\eta_\ell''}\cdot\vartheta[\eta_{\{k,\ell\}}+\eta_k](z) = e^{4\pi i({}^t\eta_k'\cdot\eta_\ell''-{}^t\eta_k''\cdot\eta_\ell')}\,\vartheta[\eta_\ell](z) = -\vartheta[\eta_\ell](z).)$$

This gives

$$D_\infty(X_kY_\ell-X_\ell Y_k) = -e^{4\pi i\,{}^t\eta_k'\cdot\eta_\ell''}\,\frac{D_\infty\vartheta[\eta_{\{k,\ell\}}](0)^2}{\vartheta[\eta_k](0)\vartheta[\eta_\ell](0)}\cdot\frac{\vartheta[\eta_k](z)\cdot\vartheta[\eta_\ell](z)}{\vartheta[0](z)^2}$$

$$= (a_k-a_\ell)e^{4\pi i\,{}^t\eta_\ell'\eta_k''}\,\frac{D_\infty\vartheta[\eta_{\{k,\ell\}}](0)\cdot D_k\vartheta[\eta_{\{k,\ell\}}](0)\vartheta[0](0)^2}{\vartheta[\eta_k](0)^2\,\vartheta[\eta_\ell](0)^2}\,X_kX_\ell\,,$$

which by Corollary 9.8 is $(a_k-a_\ell)X_kX_\ell$. <u>QED</u>

The rest of the proof is now simple. It follows that the image of ϕ is contained in one of the orbits of the g flows X_{F_k}, i.e., in one of the complex varieties

$$F_k = c_k, \qquad k \in U.$$

But these are affine pieces of 2^{k+1}-sheeted coverings Y_c of jacobians J_c of hyperelliptic curves (or of generalized jacobians of singular limits of hyperelliptic curves). Since the differential of ϕ carries the invariant vector fields on X'_Ω to the invariant vector fields of the algebraic groups Y_c, ϕ must extend to an everywhere-defined homomorphisms

$$\phi\colon X'_\Omega \longrightarrow Y_c,$$

with finite kernel. In particular, Y_c is also compact, hence is a covering of a jacobian of a smooth hyperelliptic curve. Next, the finite group

$$\mathrm{Ker}(X'_\Omega \longrightarrow X\ )$$

and the finite group

$$\mathrm{Ker}(Y_c \longrightarrow J_c)$$

both act on the coordinates x_i, y_i by sign changes $(x_i, y_i) \longmapsto (\epsilon_i x_i, \epsilon_i y_i)$, hence ϕ descends to

$$\phi_0\colon X_\Omega \longrightarrow J_c.$$

But by construction $\phi_0\big(X_\Omega - V(\vartheta[0])\big) \subset J_c - \theta$, hence $\phi^{-1}(\theta) = V(\vartheta[0])$. If ϕ_0 had a kernel, the divisor $V(\vartheta[0])$ would be invariant under a non-trivial translation, which it is not. Therefore ϕ_0 is an isomorphism of X_Ω with the jacobian J_c and $V(\vartheta[0])$ is isomorphic to θ.

QED

§10. The hyperelliptic $\wp$-function

On any hyperelliptic jacobian Jac C, there is one meromorphic function which is most important, playing a central role in the function theory on Jac C. When $g = 1$, this function is Weierstrass' $\wp$-function, so, at the risk of precipitating some confusion in notation, we want to call this function $\wp(\vec{z})$ too.

We fix a hyperelliptic curve C, and let:

B = branch points of C

$\infty \in B$

$U \subset B-\infty$ a set of $g+1$ points

t = tangent vector to C at ∞.

This defines for us

a) an invariant vector field D_∞ on Jac C. Namely, if $\{\omega_i\}$ is a basis of $\Gamma(\Omega^1)$, $z_i = \int^P \omega_i$ are coordinates on Jac C, and $\langle\omega_i(\infty),t\rangle = e_i$, then $D_\infty = \sum e_i\, \partial/\partial z_i$.

b) a definite theta divisor $\Theta \subset$ Jac C. Namely, Θ is the locus of divisor classes

$$\sum_1^{g-1} D_i - \Big(\sum_{Q\in U} Q - 2\infty\Big)$$

c) the $\wp$-function. Namely, let Θ be given by $\vartheta(\vec{z}) = 0$, then

$$\wp(\vec{z}) = D_\infty^2 \log \vartheta(\vec{z}).$$

Note that $\wp$ is L_Ω-periodic, hence is a rational function on the variety Jac C. More intrinsically, $\wp(\vec{z})$ is characterized up to an additive constant as the unique rational function f on Jac C such that

> For all $U \subset$ Jac C open,
> for all holomorphic g on U, g vanishing to order 1 on $\Theta \cap U$ and nowhere else,
> $f = D_\infty^2 \log g$ + holo. fcn. on U.

(In fact, we can even construct $\wp(\vec{z})$ in characteristic p! Start with a Zariski-open covering $\{U_\alpha\}$ of Jac C and local equations f_α of $\Theta \cap U_\alpha$ in U_α. Then f_α/f_β is a unit in $U_\alpha \cap U_\beta$, hence

$$\frac{D_\infty f_\alpha}{f_\alpha} - \frac{D_\infty f_\beta}{f_\beta}$$

is a 1-cochain in $\mathcal{O}_{\text{Jac C}}$. But $D_\infty : H^1(\mathcal{O}_J) \longrightarrow H^1(\mathcal{O}_J)$ is zero, hence

$$D_\infty \frac{D_\infty f_\alpha}{f_\alpha} - D_\infty \frac{D_\infty f_\beta}{f_\beta} = g_\alpha - g_\beta$$

where $g_\alpha \in \Gamma(U_\alpha, \mathcal{O}_{\text{Jac C}})$. Let

$$\wp = D_\infty\left(\frac{D_\infty f_\alpha}{f_\alpha}\right) - g_\alpha . \Big)$$

How much does $\wp$ depend on the given data? $1^{\underline{st}}$, the additive constant in $\wp$ depends on the choice of ϑ itself, i.e., the choice of homology basis $\{A_i\},\{B_i\}$. If t is changed, $\wp$ will be replaced by $c\cdot\wp$. If U is changed, $\wp(\vec{z})$ is replaced by $\wp(\vec{z}+\vec{a})$, $\vec{a} \in \text{Jac } C_2$. Thus $\wp$ really depends essentially only on C and ∞, though to get a definite $\wp$ many further choices must be made. Note that $\wp(-\vec{z}) = \wp(\vec{z})$. We can easily identify $\wp$ in our affine model. Let C be given by $s^2 = f(t) = \prod_1^{2g+1} (t-a_i)$.

Proposition 10.1. In the affine model of Jac C:

$$\text{Jac } C - \Theta = \{(U,V,W) \mid f-V^2 = U\cdot W, \text{ degrees as before}\}$$

let

$$U(t) = t^g + U_1 t^{g-1} + \cdots$$

$$W(t) = t^{g+1} + W_0 t^g + \cdots .$$

Note that $U_1 + W_0 = -\sum_{i=1}^{2g+1} a_i$. Assume t chosen so that $D_\infty \mathbf{U} = V$. Then

$$\frac{U_1 - W_0}{2} = \frac{1}{2}(\sum a_k x_k^2 - \sum y_k^2) = 4\wp(\vec{z}) + d$$

for some constant d.

Proof: Recall that $f_1(t) = \prod_{i \in U} (t-a_i)$ and

$$\frac{U(t)}{f_1(t)} = \sum_{k \in U} \frac{x_k^2}{t-a_k}, \quad \frac{V(t)}{f_1(t)} = \sqrt{-1} \cdot \sum_{k \in U} \frac{x_k y_k}{t-a_k}, \quad \frac{W(t)}{f_1(t)} = 1 + \sum_{k \in U} \frac{y_k^2}{t-a_k}$$

If we expand

$$U(t) = \sum_k \prod_{\ell \neq k} (t-a_\ell) \cdot x_k^2$$

$$= \left(\sum x_k^2\right) t^g + \left(\sum_k a_k x_k^2 - \sum_\ell a_\ell \cdot \sum x_k^2\right) t^{g-1} + \cdots\cdots$$

we see that

$$U_1 = \sum a_k x_k^2 - \sum a_k .$$

Similarly,

$$W(t) = \prod_{\ell} (t-a_\ell) + \sum_k \prod_{\ell \neq k} (t-a_\ell) y_k^2$$

$$= t^{g+1} + \left(\sum y_k^2 - \sum a_k\right) t^g + \cdots .$$

hence

$$W_0 = \sum y_k^2 - \sum a_k .$$

Therefore

$$U_1 - W_0 = \sum a_k x_k^2 - \sum y_k^2,$$

which proves the 1st equality.

Now $D_\infty U = V$, so we find

$$D_\infty(x_k^2) = \sqrt{-1}\, x_k y_k$$

or

$$D_\infty x_k = \frac{\sqrt{-1}}{2} y_k ,$$

i.e., $D_\infty = \frac{\sqrt{-1}}{2} \times$ (the derivative of Neumann's dynamical system).

Now in Neumann's system

$$\ddot{x}_k = -a_k x_k + x_k \left(\sum_\ell a_\ell x_\ell^2 - \sum_\ell y_\ell^2\right) ,$$

hence

$$D_\infty^2(x_k) = -\frac{1}{4}\left(-a_k + \sum a_\ell x_\ell^2 - \sum y_\ell^2\right) x_k .$$

Now in terms of theta functions

$$x_k = c_k \frac{\vartheta[\eta_k]}{\vartheta[0]} \quad .$$

Consider the difference

$$D^2(\log\vartheta[0]) - \frac{1}{8}\Big(\sum_{\ell\in U} a_\ell x^2 - \sum_{\ell\in U} y_\ell^{\,2}\Big) \quad .$$

The first term equals

$$\frac{D_\infty^{\,2}\vartheta[0]}{\vartheta[0]} - \left(\frac{D_\infty\vartheta[0]}{\vartheta[0]}\right)^2$$

and the second equals

$$+\frac{1}{2}\,\frac{D_\infty^2 x_k}{x_k} - \frac{1}{8}\,a_k \,.$$

Working this out,

$$D_\infty^2(\log\vartheta[0]) - \frac{1}{8}\Big(\sum a_\ell x_\ell^2 - \sum y_\ell^2\Big) = \frac{1}{2}\,\frac{D_\infty^2\vartheta[\eta_k]}{\vartheta[\eta_k]} - \frac{D_\infty\vartheta[\eta_k]\cdot D_\infty\vartheta[0]}{\vartheta[\eta_k]\cdot\vartheta[0]} + \frac{1}{2}\,\frac{D_\infty^2\vartheta[0]}{\vartheta[0]} - \frac{1}{8}\,a_k$$

which has at most simples poles at $V(\vartheta[0]) \cup V(\vartheta[\eta_k])$. Since this is true for every k, the difference has in fact only poles at $V(\vartheta[0])$. But the only functions with only simple poles at $V(\vartheta[0])$ are constants, and this proves the second equality in the Proposition. QED

More generally, we can relate all the functions on Jac C defined by the coefficients of U,V on the one hand, and by derivatives of $\log\vartheta[0](z)$ on the other:

Proposition 10.2. (I) The two vector spaces of rational functions on Jac C spanned by

a) $D D_\infty \log \vartheta[0](z)$, all invariant vector fields D, and 1

b) the coefficients U_i of $U(t)$ including $U_0 = 1$

are equal. This space has dimension g+1, and consists of even functions with at most double poles at Θ.

(II) Likewise, the two vector spaces of rational functions on Jac C spanned by

a) $D D_\infty^2 \log \vartheta[0](z)$, all invariant vector fields D

b) the coefficients V_i of $V(t)$

are equal. This space has dimension g, and consists of odd functions with poles of order exactly three on Θ.

In fact, for suitable constants c,c' and d_k,

$$D_{a_k} D_\infty \log \vartheta[0](z) = c\,\lambda_k^D + d_k$$

$$D_{a_k} D_\infty^2 \log \vartheta[0](z) = c\,\mu_k^D \qquad \text{, for all } k \in U.$$

Proof: We calculate $D_{a_k} D_\infty^2 \log \vartheta[0]$ as follows:

$$D_{a_k} D_\infty^2 \log \vartheta[0] = \frac{1}{8} D_{a_k}(U_1 - W_0) \qquad \text{by Prop. 10.1}$$

$$= \frac{1}{4} D_{a_k}(U_1) - \frac{1}{8} D_{a_k}(U_1 + W_0).$$

But $U_1 + W_0 = -\sum_1^{2g+1} a_i$ is constant on Jac C, and

$$D_{a_k} U = \frac{V(a_k)U(t) - U(a_k)V(t)}{t - a_k} \, ,$$

hence

$$D_{a_k} U_1 = V(a_k) = c_1 \mu_k$$

for some constant c_1 depending only on C. This proves

$$D_{a_k} D_\infty^2 \log \vartheta[0] = c\, \mu_k$$

and hence proves (II). Moreover, as $D_{a_k} \lambda_k = \mu_k$, it proves that $f_k = D_{a_k} D_\infty \log \vartheta[0] - c\lambda_k$ is a function on Jac C killed by D_∞. But f_k has poles only on Θ and either f_k is a constant or D_∞ must be everywhere tangent to Θ. As this latter is not the case, f_k is a constant, which proves (I). QED

We now come to the main point of this section: we ask whether we can coordinatize Jac C by using the function $\wp(z)$ and its derivatives along D_∞:

$$\wp^{(k)}(z) \underset{\text{def}}{=} D_\infty^k \wp(z)$$

only. The fact that this is possible was discovered by McKean-Van Moerbeke in their beautiful paper*. Not only is this possible, but this leads to an affine embedding of Jac C-Θ governed by a quite intricate algebra.

To be precise, we fix n and consider the morphism

$$\phi_n : \text{Jac } C - \Theta \longrightarrow \mathbb{C}^n$$
$$z \longmapsto (\wp(z), \wp^{(1)}(z), \cdots, \wp^{(n-1)}(z)) \, .$$

*The spectrum of Hill's equation, Inv. Math., 30, 1975

Theorem 10.3. If $n = 2g$, ϕ_n is an embedding, hence $\wp^{(i)}(z)$, $0 \le i \le 2g-1$, generate the affine ring of Jac $C - \Theta$. In fact, we may solve for U_i, V_i, W_i in terms of $\wp^{(k)}$ and a_k, and for $\wp^{(k)}$ in terms of U_i, V_i, W_i and a_k by means of "universal polynomials".

Proof: We shall not find the formulae relating the $\{U_i, V_i, W_i\}$ and $\{\wp^{(k)}\}$ directly, but rather via a third set of variables $\{U_i^*, V_i^*, W_i^*\}$. Our first job is to introduce these. We convert the identity

$$\text{a)} \qquad f = UW + V^2$$

between polynomials in t to an identity between polynomials in t^{-1}:

$$\text{b)} \qquad \frac{f(t)}{t^{2g+1}} = \left(\frac{U(t)}{t^g}\right)\left(\frac{W(t)}{t^{g+1}}\right) + \frac{1}{t}\left(\frac{V(t)}{t^g}\right)^2 .$$

A polynomial in t^{-1}, with constant term 1, has a unique square root in the ring of power series, with constant term 1, so we write

$$\frac{f(t)}{t^{2g+1}} = \prod_{i=1}^{2g+1} (1-a_i t^{-1}) = (1 + \alpha_1 t^{-1} + \alpha_2 t^{-2} + \cdots\cdots)^2$$

for suitable constants $\alpha_1, \alpha_2, \cdots$ and write

$$\phi(t^{-1}) = 1 + \alpha_1 t^{-1} + \alpha_2 t^{-2} + \cdots\cdots\cdots .$$

Thus (b) can be written:

$$\text{c)} \qquad 1 = \left(\frac{U(t)\cdot t^{-g}}{\phi(t^{-1})}\right) \cdot \left(\frac{W(t)\cdot t^{-g-1}}{\phi(t^{-1})}\right) + t^{-1}\left(\frac{V(t)\cdot t^{-g}}{\phi(t^{-1})}\right)^2 .$$

Let $U^*(t^{-1}) \underset{\text{def}}{=} \dfrac{U(t)\cdot t^{-g}}{\phi(t^{-1})}$

$$= 1+U_1^*t^{-1}+U_2^*t^{-2}+\cdots\cdots$$

$$V^*(t^{-1}) \underset{\text{def}}{=} \frac{V(t)\cdot t^{-g}}{\phi(t^{-1})}$$

$$= V_1^*t^{-1}+V_2^*t^{-2}+\cdots\cdots\cdot$$

$$W^*(t^{-1}) \underset{\text{def}}{=} \frac{W(t)\cdot t^{-g-1}}{\phi(t^{-1})}$$

$$= 1 + W_0^*t^{-1}+W_1^*t^{-2}+\cdots\cdots\cdot$$

so that c) becomes

$$\text{d)} \quad 1 = U^*(t^{-1})\cdot W^*(t^{-1})+t^{-1}\cdot V^*(t^{-1})^2 \ .$$

Note that the (U_i^*,V_i^*,W_i^*) and the (U_i,V_i,W_i) determine each other given the α_i, by the universal polynomials obtained by equating coefficients of t^{-n} in:

$$(1+U_1t^{-1}+U_2t^{-2}+\cdots) = (1+U_1^*t^{-1}+U_2^*t^{-1}+\cdots)\cdot(1+\alpha_1t^{-1}+\alpha_2t^{-2}+\cdots)$$

$$\text{e)} \quad (\ V_1t^{-1}+V_2t^{-2}+\cdots) = (\ V_1^*t^{-1}+V_2^*t^{-1}+\cdots)\cdot(1+\alpha_1t^{-1}+\alpha_2t^{-2}+\cdots)$$

$$(1+W_0t^{-1}+W_1t^{-2}+\cdots) = (1+W_0^*t^{-1}+W_1^*t^{-2}+\cdots)\cdot(1+\alpha_1t^{-1}+\alpha_2t^{-2}+\cdots)$$

e.g.,

$$U_1 = U_1^*+\alpha_1$$

$$V_1 = V_1^*$$

$$W_0 = W_0^*+\alpha_1 \ .$$

On the other hand, (d) written out gives a recursive procedure for finding the W_i^* from (U_i^*,V_i^*), viz.

$$U_1^*+W_0^* = 0$$

$$U_2^*+U_1^*W_0^*+W_1^* = 0$$

f) $$U_3^*+U_2^*W_0^*+U_1^*W_1^*+W_2^*+V_1^{*2} = 0$$

$$\cdots\cdots$$

$$U_n^*+W_{n-1}^*+\begin{pmatrix}\text{univ. polyn. in } U_1^*,\cdots,U_{n-1}^* \\ W_0^*,\cdots,W_{n-2}^*,V_1^*,\cdots,V_{n-2}^*\end{pmatrix} = 0$$

Note that

$$\wp = \tfrac{1}{8}(U_1-W_0) = \tfrac{1}{8}(U_1^*-W_0^*) = \tfrac{1}{4}U_1^*.$$

The flow D_∞ can be easily written in terms of U^*,V^*,W^*. It comes out as

$$\dot{U}^* = V^*$$

$$\dot{V}^* = \tfrac{1}{2}t(-W^*+(1-8\wp\cdot t^{-1})U^*)$$

$$\dot{W}^* = -(1-8\wp\cdot t^{-1})V^*$$

or

$$\dot{U}_i^* = V_i^*$$

g) $$\dot{V}_i^* = \tfrac{1}{2}(-W_i^*+U_{i+1}^*-2U_1^*\cdot U_i^*)$$

$$\dot{W}_i = -V_{i+1}^*+2U_1^*\cdot V_i^*$$

These give us, by induction, the formulae:

h)

$$4\wp = U_1^*$$
$$4\wp^{(1)} = \dot{U}_1^* = V_1^*$$
$$4\wp^{(2)} = \dot{V}_1^* = \tfrac{1}{2}(-W_1^*+U_2^*-2U_1^{*2})$$
$$= U_2^* - \tfrac{3}{2}U_1^{*2} \qquad \text{(using } W_1^* = -U_2^*+U_1^{*2})$$
$$4\wp^{(3)} = \dot{U}_2^* - 3U_1^*\dot{U}_1^*$$
$$= V_2^* - 3U_1^*V_1^*$$
$$\cdots\cdots\cdots\cdots$$
$$\wp^{(2k)} = U_{k+1}^* + (\text{polyn. in } U_1^*,\cdots,U_k^*,V_1^*,\cdots,V_{k-1}^*)$$
$$\wp^{(2k+1)} = V_{k+1}^* + (\text{polyn. in } U_1^*,\cdots,U_{k+1}^*,V_1^*,\cdots,V_k^*) \quad .$$

We may solve these backwards:

i)

$$U_1^* = 4\wp$$
$$V_1^* = 4\wp^{(1)}$$
$$U_2^* = 4\wp^{(2)}+24\wp^2$$
$$V_2^* = 4\wp^{(3)}+48\,\wp\cdot\wp^{(1)}$$
$$\cdots\cdots\cdots\cdots$$
$$U_{k+1}^* = \wp^{(2k)}+\left(\text{polyn. in } \wp,\wp^{(1)},\cdots,\wp^{(2k-1)}\right), \text{ call this } F_{k+1}(\wp,\wp^{(1)},\cdots,\wp^{(2k)})$$
$$V_{k+1}^* = \wp^{(2k+1)}+\left(\text{polyn. in } \wp,\wp^{(1)},\cdots,\wp^{(2k)}\right), \text{ call this } G_{k+1}(\wp,\wp^{(1)},\cdots,\wp^{(2k+1)}).$$

It is easy to set up a recursive procedure which determines the sequences $\{F_k\},\{G_k\}$. First of all, as

$$V_k^* = \dot{U}_k^*,$$

it follows

$$G_k(\wp,\cdots,\wp^{(2k-1)}) = \dot{F}_k(\wp,\cdots,\wp^{(2k-2)}) \tag{10.4}$$

The dot here means this: if $F(\wp,\wp^{(1)},\cdots,\wp^{(n)})$ is any polynomial, then:

$$\dot{F}(\wp,\wp^{(1)},\cdots,\wp^{(n+1)}) = \sum_{k=0}^{n} \frac{\partial F}{\partial \wp^{(k)}} \cdot \wp^{(k+1)} .$$

Moreover:

$$\begin{aligned}
\dddot{U}_k^* &= \ddot{V}_k^* \\
&= \frac{1}{2}(-W_k^* + U_{k+1}^* - 8\wp\cdot U_k^*)^{\bullet} \\
&= \frac{1}{2}((V_{k+1}^*-8\wp V_k^*) + V_{k+1}^* - 8(\wp U_k^*)^{\bullet}) \\
&= V_{k+1}^* - 4\wp\cdot V_k^* - 4(\wp U_k^*)^{\bullet} ,
\end{aligned}$$

hence

$$\dddot{F}_k + 4\wp\cdot\dot{F}_k + 4(\wp\cdot F_k)^{\bullet} = G_{k+1} = \dot{F}_{k+1} \tag{10.5}$$

Then (10.4) and (10.5) determine the polynomials $\{F_k\},\{G_k\}$, given the extra facts that F_k,G_k have no constant terms and that the map

$$\cdot: \mathbb{C}[\wp,\wp^{(1)},\wp^{(2)},\cdots] \longrightarrow \mathbb{C}[\wp,\wp^{(1)},\wp^{(2)},\cdots]$$

has no kernel except for constants. We also note for future use that W_i^* is given in terms of the $\wp^{(k)}$ by the $2^{\underline{nd}}$ equation in (g):

$$\begin{aligned} W_i^* &= U_{i+1}^* - 2\dot{V}_i^* - 8\wp \cdot U_i^* \\ &= F_{i+1} - 2\dot{G}_i - 8\wp \cdot F_i \quad . \end{aligned}$$

Algebraically, we have shown that the 2 polynomial rings $R_{U,V,W}^* = \mathbb{C}[U_1^*,U_2^*,\cdots;V_1^*,V_2^*,\cdots;W_0^*,W_1^*,\cdots] \,/\text{identities (f)}$

and

$$R_\wp = \mathbb{C}[\wp,\wp^{(1)},\wp^{(2)},\cdots]$$

are isomorphic, by an isomorphism that carries the derivation of $R_{U,V,W}^*$ defined by (g) to the derivation of $R_\wp$ given by $\dot{\wp}^{(k)} = \wp^{(k+1)}$, and carrying the subring

$$R_{U,V,W}^{*,g} = \mathbb{C}[U_1^*,\cdots,U_g^*,V_1^*,\cdots,V_g^*,W_0^*,\cdots,W_{g-1}^*] \Big/ \begin{pmatrix}\text{First g identities}\\ \text{in (f)}\end{pmatrix}$$

to the subring

$$R_\wp^g = \mathbb{C}[\wp,\wp^{(1)},\cdots,\wp^{(2g-1)}].$$

To finish the proof of the Theorem, note that by (e), the functions $U_1,\cdots,U_g,V_1,\cdots,V_g$ and hence the whole affine ring of Jac C-Θ are polynomials in $U_1^*,\cdots,U_g^*,V_1^*,\cdots,V_g^*$, hence by what we have just said, polynomials $\wp$, $\wp^{(1)},\cdots,\wp^{(2g-1)}$. Thus ϕ_{2g} is an embedding. <u>QED</u>

Still imitating the algebra of McKean and Van Moerbeke, we can go further and explicitly describe the equations in $\wp, \wp^{(1)}, \cdots, \wp^{(2g)}$ that define $\phi_{2g+1}(\text{Jac } C-\Theta) \subset \mathbb{C}^{2g+1}$. The result is this:

Theorem 10.6: I) There are unique polynomials without constant term

$$H_{k\ell} \in \mathbb{C}[\wp, \wp^{(1)}, \wp^{(2)}, \cdots]$$

such that

$$\dot{H}_{k\ell} = G_k \cdot F_\ell.$$

In fact,

$$H_{k,\ell} \in \mathbb{C}[\wp, \cdots, \wp^{(n)}], \ n = \max(2k-2, 2\ell-3), \quad \text{and} \quad H_{k,0} = F_k.$$

II) If $\phi(t^{-1}) = 1 + \alpha_1 t^{-1} + \alpha_2 t^{-2} + \cdots$, then $\phi_{2g+1}(\text{Jac } C-\Theta)$ is defned by

$$\sum_{k=1}^{g+1} \alpha_{g+1-k} H_{k\ell} = -\alpha_{\ell+1+g}, \quad 0 \leq \ell \leq g.$$

Proof: The method we follow is the most direct one, but it unfortunately requires a rather nasty computation at the end. If $F(t^{-1}) = a_0 + a_1 t^{-1} + \cdots$ is any power series, write

$$\delta F = a_{g+1} + a_{g+2} t^{-1} + \cdots\cdots$$

for the "tail" of F starting at the t^{-g-1}-terms, and let

$$\tilde{F} = a_0 + a_1 t^{-1} + \cdots + a_g t^{-g}$$

be the "head" of F. Thus

$$F = \tilde{F} + t^{-g-1}.\delta F.$$

Now

$$\frac{U(t)}{t^g} = \phi(t^{-1})U^*(t^{-1})$$

$$= (\tilde{\phi}+t^{-g-1}.\delta\phi)\cdot(\tilde{U}^*+t^{-g-1}.\delta U^*)$$

is a polynomial of degree g in t^{-1}. For simplicity, we drop the * and write $\tilde{U}, \delta U$ for $\tilde{U}^*, \delta U^*$, and similarly for $\tilde{V}, \delta V, \tilde{W}, \delta W$ below. It follows:

$$0 = \delta\left[(\tilde{\phi}+t^{-g-1}\delta\phi)\cdot(\tilde{U}+t^{-g-1}.\delta U)\right]$$

$$\equiv \delta(\tilde{\phi}\cdot\tilde{U}) + \tilde{\phi}\cdot\delta U + \delta\phi\cdot\tilde{U} \mod t^{-g-1},$$

hence

a) $\delta U \equiv -\tilde{\phi}^{-1}.\left(\delta(\tilde{\phi}\cdot\tilde{U}) + \delta\phi\cdot\tilde{U}\right) \mod t^{-g-1}.$

This formula enables one to solve for the terms in U* between t^{-g-1} and t^{-2g-1} using the terms between 1 and t^{-g}, given that U* comes from a polynomial U in t of degree g. Similarly, we get formulae

b) $\delta V \equiv -\tilde{\phi}^{-1}.\left(\delta(\tilde{\phi}\cdot\tilde{V}) + \delta\phi\cdot\tilde{V}\right) \mod t^{-g-1}$

c) $\delta W \equiv -\tilde{\phi}^{-1}.\left(\delta(\tilde{\phi}\cdot\tilde{W}) + \delta\phi\cdot\tilde{W}-W_g\right) \mod t^{-g-1}.$

(In (c), the fact that W has degree g+1 makes the formula have an extra term.)

Now start with values of $\wp, \wp^{(1)}, \cdots, \wp^{(2g)}$ and ask whether they give a point of $\phi_{2g+1}(\text{Jac } C-\Theta)$. From these values of $\wp^{(i)}$, we define the numbers

$$U_1^*,\cdots,U_g^*,V_1^*,\cdots,V_g^*,W_0^*,\cdots,W_g^*$$

by the universal polynomials of Theorem 10.3, hence the polynomials $\hat{U},\tilde{V}$ and $\tilde{W}$ as well as the one extra number W_g^*. These in turn define unique polynomials $U(t),V(t),W(t)$ such that

$$\widetilde{\left(\frac{U(t)t^{-g}}{\phi(t^{-1})}\right)} \equiv \tilde{U} \bmod t^{-g-1}$$

$$\widetilde{\left(\frac{V(t)t^{-g}}{\phi(t^{-1})}\right)} \equiv \tilde{V} \bmod t^{-g-1}$$

$$\widetilde{\left(\frac{W(t)t^{-g-1}}{\phi(t^{-1})}\right)} \equiv \tilde{W} + W_g^* t^{-g-1} \bmod t^{-g-2}.$$

The condition that we have a point of Im ϕ_{2g+1} is that

$d_1)$ $\quad f = UW+V^2$.

But we can rewrite this condition as

$$d_2) \quad 1 \equiv \left(\frac{U(t)t^{-g}}{\phi(t^{-1})}\right)\left(\frac{W(t)t^{-g-1}}{\phi(t^{-1})}\right) + t^{-1}\left(\frac{V(t)t^{-g}}{\phi(t^{-1})}\right)^2 \bmod t^{-2g-2} .$$

Now, mod t^{-2g-2}, we have seen that

$$\frac{U(t)t^{-g}}{\phi(t^{-1})} \equiv \tilde{U} - t^{-g-1}\tilde{\phi}^{-1}\cdot\left(\delta(\tilde{\phi}\tilde{U})+\delta\phi\cdot\tilde{U}\right) \bmod t^{-2g-2}$$

$$\frac{V(t)t^{-g}}{\phi(t^{-1})} \equiv \tilde{V} - t^{-g-1}\tilde{\phi}^{-1}\cdot\left(\delta(\tilde{\phi}\tilde{V})+\delta\phi\cdot\tilde{V}\right) \bmod t^{-2g-2}$$

$$\frac{W(t)t^{-g-1}}{\phi(t^{-1})} \equiv \tilde{W} - t^{-g-1}\tilde{\phi}^{-1}\cdot\left(\delta(\tilde{\phi}\tilde{W})+\delta\phi\cdot\tilde{W}-W_g\right) \bmod t^{-2g-2} .$$

Therefore equation (d_2) is equivalent to

$$(d_3)\quad 1 \equiv \tilde{U}\cdot\tilde{W}+t^{-1}\tilde{V}^2-t^{-g-1}\tilde{\phi}^{-1}\Big[\tilde{W}\cdot\delta(\tilde{\phi}\cdot\tilde{U})+2\tilde{W}\cdot\tilde{U}\cdot\delta\phi+\tilde{U}\delta(\tilde{\phi}\cdot\tilde{W})$$
$$-\tilde{U}\cdot W_g+2t^{-1}\tilde{V}\cdot\delta(\tilde{\phi}\tilde{V})+2t^{-1}\tilde{V}^2\cdot\delta\phi\Big] \bmod t^{-2g-2} \ .$$

As the terms in $t^0, t^{-1}, \cdots, t^{-g}$ cancel automatically by definition of the universal polynomials for the W^*_ℓ, this reduces to

$$(d_4)\quad \tilde{\phi}\cdot\delta\left(\tilde{U}\cdot\tilde{W}+t^{-1}\tilde{V}^2\right) \equiv 2\delta\phi+\tilde{W}\cdot\delta(\tilde{\phi}\cdot\tilde{U})+\tilde{U}(\delta(\tilde{\phi}\cdot\tilde{W})-W_g)$$
$$+2t^{-1}\tilde{V}\cdot\delta(\tilde{\phi}\cdot\tilde{V}) \bmod t^{-g-1} \ .$$

First look at the constant terms in this equation.
To calculate this, note that $\delta(U^*\cdot W^*+t^{-1}V^{*2}) = 0$, hence

$$\begin{aligned}&\text{constant term in } \delta(\tilde{U}\cdot\tilde{W}+t^{-1}\tilde{V}^2)\\ &\qquad = -\text{constant term in } \delta(U^*W^*-\tilde{U}\cdot\tilde{W}+t^{-1}(V^{*2}-\tilde{V}^2))\\ &\qquad = -(U^*_{g+1}+W^*_g) \ .\end{aligned}$$

Altogether the constant terms give us:

$$-(U^*_{g+1}+W^*_g) = 2\alpha_{g+1}+(\alpha_1U^*_g+\cdots+\alpha_gU^*_1)+(\alpha_1W^*_{g-1}+\cdots+\alpha_gW^*_0)-W_g .$$

Since $W_g = W^*_g+\alpha_1W^*_{g-1}+\cdots+\alpha_gW^*_0+\alpha_{g+1}$, this reduces to

$$(e_1)\qquad -\alpha_{g+1} = U^*_{g+1} + \alpha_1U^*_g + \cdots + \alpha_gU^*_1$$

which is to be the equation in (II) for $\ell = 0$, i.e., set $H_{k,0} = F_k$ and then (e_1) is:

$$(e_2)\qquad \sum_{k=1}^{g+1} \alpha_{g+1-k}H_{k,0} = -\alpha_{g+1} \ .$$

To get the remaining equations, substitute into (d_4)

$$W_g = W_g^* + \alpha_1 W_{g-k}^* + \cdots + \alpha_g W_0^* - U_{g+1}^* - \alpha_1 U_g^* - \cdots - \alpha_g U_1^*$$

and write (d_4) as

$$(f_1) \quad -\delta\phi \equiv \tfrac{1}{2}\tilde{W}\delta(\tilde{\phi}\tilde{U}) + \tfrac{1}{2}\tilde{U}(\delta(\tilde{\phi}\cdot\tilde{W}) - (W_g^* + \cdots + \alpha_g W_0^* - U_{g+1}^* - \cdots - \alpha_g U_1^*))$$
$$+ t^{-1}\tilde{V}\delta(\tilde{\phi}\tilde{V}) - \tfrac{1}{2}\tilde{\phi}\delta(\tilde{U}\tilde{W} + t^{-1}\tilde{V}^2) \bmod t^{-2g-2} .$$

Expand this into

$$(f_2) \qquad - \sum_{\ell=0}^{g} \alpha_{\ell+1+g} t^{-\ell} = \sum_{\ell=0}^{g} \sum_{k=1}^{g+1} \alpha_{g+1-k} H_{k\ell}(\wp, \wp', \ldots, \wp^{(2g)}) \cdot t^{-\ell}$$

so that the coefficients of each $t^{-\ell}$ give us the remaining equations in the form required. It remains to check that $\dot{H}_{k\ell} = G_k \cdot F_\ell$. This should have a conceptual proof but it is not too hard to check by directly differentiating.

We use D for $\dot{}$ in this calculation.

Thus to start, note that

$$\begin{aligned} D(\tilde{U}) &= \tilde{V} \\ 2D(\tilde{V}) &= -t(\tilde{W} + t^{-g-1}W_g^*) + t(\tilde{U} + t^{-g-1}U_{g+1}^*) - 8\wp\cdot\tilde{U} \\ D(\tilde{W}) &= -\tilde{V} + 8\wp\cdot t^{-1}(\tilde{V} - V_g^* t^{-g}) \end{aligned}$$

from which one deduces

$$D(\tilde{U}\cdot\tilde{W}) + t^{-1}\tilde{V}^2) = -8\wp\tilde{U}\cdot V_g^* t^{-g-1} - \tilde{V}\cdot W_g^* t^{-g-1} + \tilde{V}U_{g+1}^* t^{-g-1}.$$

Likewise, one has

$$D(\tilde{W}\cdot\delta(\widetilde{\phi U})) = (-\tilde{V}+8\wp t^{-1}(\tilde{V}-V_g^* t^{-g}))\delta(\widetilde{\phi U}) + \tilde{W}\cdot\delta(\widetilde{\phi V})$$

$$D(\tilde{U}\cdot\delta(\widetilde{\phi W})) = \tilde{V}\cdot\delta(\widetilde{\phi W}) + \tilde{U}\cdot\delta[\tilde{\phi}(-\tilde{V}+8\wp t^{-1}(\tilde{V}-V_g^* t^{-g}))]$$

$$D(2t^{-1}\tilde{V}\cdot\delta(\tilde{\phi}\tilde{V})) = (-(\tilde{W}+t^{-g-1}W_g^*) + (\tilde{U}+t^{-g-1}U_{g+1}^*) - 8\wp t^{-1}\tilde{U})\delta(\tilde{\phi}\tilde{V})$$

$$+t^{-1}\tilde{V}.\delta\Big[\tilde{\phi}(-t(\tilde{W}+t^{-g-1}W_g^*)+t(\tilde{U}+t^{-g-1}U_{g+1}^*)-8\wp\tilde{U})\Big] .$$

Adding these up, we get a lot of cancellation, leading to

$$D\{\tilde{\phi}\cdot\ \delta(\tilde{U}\tilde{W}+t^{-1}\tilde{V}^2)-\tilde{W}\cdot\delta(\widetilde{\phi U})-\tilde{U}\cdot\delta(\tilde{\phi}\cdot\tilde{W})-2t^{-1}\tilde{V}\cdot\delta(\tilde{\phi}\tilde{V})+\tilde{U}\cdot W_g\}$$

$$= \tilde{V}\cdot(U_{g+1}^*-W_g^*)-8\wp\tilde{U}[\delta(t^{-1}\widetilde{\phi V})-t^{-1}\delta(\tilde{\phi}\cdot\tilde{V})]$$

$$+ \tilde{V}[\delta(\widetilde{\phi U})-t^{-1}\delta(t\widetilde{\phi U})]-\tilde{V}[\delta(\widetilde{\phi W})-t^{-1}\delta(t\widetilde{\phi W})]$$

$$+ \tilde{V}\cdot(W_g^*+\alpha_1 W_{g-1}^*+\cdots+\alpha_{g+1}) + \tilde{U}\cdot(-V_{g+1}+8\wp V_g)$$

$$= \tilde{V}\cdot(U_{g+1}^*+\alpha_1 U_g^*+\cdots+\alpha_{g+1}) - \tilde{U}\cdot(V_{g+1}^*+\alpha_1 V_g^*+\cdots+\alpha_g V_1^*)$$

$$= \sum_{\ell=0}^{g}\sum_{k=0}^{g+1} \alpha_{g+1-k}(V_\ell^* U_k^*-V_k^* U_\ell^*)\cdot t^{-\ell} .$$

Thus

$$D(-2H_{k\ell}+U_k^* U_\ell^*) = V_\ell^* U_k^* - V_k^* U_\ell^*$$

which proves

$$D(H_{k\ell}) = U_\ell^* V_k^* = F_\ell\cdot G_k . \qquad \underline{\text{QED}}$$

The cases g = 1,2 are given explicitly in the table below. Note that the equation $\ell = 0$ gives $\wp^{(2g)}$ as polynomial in the lower derivatives, so that substituting this, we have exactly g equations for $\phi_{2g}(\text{Jac } C-\Theta)$ in $\mathbb{C}^{2g}$. For g = 1, this is the usual cubic equation in $\wp, \wp'$. Moreover, higher derivatives

Tables

$F_0 = 1$

$F_1 = 4\wp$

$F_2 = 4\wp''+24\wp^2$

$F_3 = 4\wp^{iv}+40(\wp')^2+80\wp\cdot\wp''+160\wp^3$

$G_0 = 0$

$G_1 = 4\wp'$

$G_2 = 4\wp'''+48\wp\cdot\wp'$

$G_3 = 4\wp^{v}+160\wp'\cdot\wp''+80\wp\cdot\wp'''+480\wp^2\cdot\wp'$

$H_{10} = 4\wp$

$H_{20} = 4\wp''+24\wp^2$

$H_{30} = 4\wp^{iv}+40(\wp')^2+80\wp\cdot\wp''+160\wp^3$

$H_{1,1} = 8\wp^2$

$H_{2,1} = 16\wp\cdot\wp''-8(\wp')^2+64\wp^3$

$H_{3,1} = 16\wp\cdot\wp^{iv}-16\wp'\cdot\wp'''+8(\wp'')^2+320\wp^2\cdot\wp''+480\wp^4$

$H_{1,2} = 8(\wp')^2+32\wp^3$

$H_{2\,2} = 8(\wp'')^2+96\wp^2\cdot\wp''+288\wp^4$

$$H_{3\,2} = 16\wp''\wp^{iv}-8(\wp''')^2+96\wp^2\cdot\wp^{iv}-192\wp\wp'\wp'''+256\wp\cdot(\wp'')^2+192(\wp')^2\wp'' +1920\wp^3\cdot\wp''+2304\wp^5 .$$

Curve of genus in 1 in $\mathbb{C}^3$ embedded by $\wp,\wp',\wp''$:

$$-\alpha_2 = \alpha_1\cdot(4\wp)+(4\wp''+24\wp^2)$$

$$-\alpha_3 = \alpha_1\cdot(8\wp^2)+(16\wp\cdot\wp''-8(\wp')^2+64\wp^3) .$$

Abelian surface in $\mathbb{C}^5$ embedded by $\wp,\wp',\wp'',\wp''',\wp^{iv}$:

$$-\alpha_3 = \alpha_2\cdot(4\wp)+\alpha_1\cdot(4\wp''+24\wp^2) + (4\wp^{iv}+40(\wp')^2+80\wp\wp''+160\wp^3)$$

$$-\alpha_4 = \alpha_2(8\wp^2)+\alpha_1(16\wp\wp''-8(\wp')^2+64\wp^3)+(16\wp\cdot\wp^{iv}-16\wp'\cdot\wp'''+8(\wp'')^2+320\wp^2\wp''+480\wp^4)$$

$$-\alpha_5 = \alpha_2(8(\wp')^2+32\wp^3)+\alpha_1(8\wp'')^2+96\wp^2\wp''+288\wp^4) + 16\wp''\wp^{iv}-8(\wp''')^2+96\wp^2\cdot\wp^{iv}-192\wp\cdot\wp'\wp'''+256\wp\cdot(\wp'')^2 + 192(\wp')^2\wp''+1920\wp^3\cdot\wp''+2304\wp^5$$

$\wp^{(n)}$, $n > 2g$, are expressed recursively in $\wp,\cdots,\wp^{(2g-1)}$ by repeated differentiation of the equation $\ell = 0$, and substitution of previous expressions for $\wp^{(m)}$, $2g \leq m < n$. Likewise, the other vector fields on Jac C can be given by elementary explicit formulae. We sketch this.

We use the basis D_k, $1 \leq k \leq g$, introduced in §3, i.e.,

$$D_P = \sum_{k=1}^{g} a^{g-k} \cdot D_k$$

for all $P \in C-(\infty)$, $a = t(P)$. Here D_1 is the D_∞ we have been working with. We showed in §3 that

$$D_k U_\ell = \sum_{\substack{i+j=k+\ell-1 \\ i \geq \max(k,\ell) \\ j \leq \min(k,\ell)-1}} (V_i U_j - V_j U_i)$$

Thus in the sum for $D_k U_1$ we have only the one term $j = 0$, $i = k$, and

$$\begin{aligned} D_k U_1 &= V_k U_0 - V_0 U_k \\ &= V_k \quad . \end{aligned}$$

Therefore

$$(10.7) \qquad \begin{aligned} 4(D_k \wp) &= D_k U_1 \\ &= V_k \\ &= V_k^* + \alpha_1 V_{k-1}^* + \cdots + \alpha_{k-1} V_1^* \\ &= \sum_{\ell=1}^{k} G_\ell(\wp, \wp^{(1)}, \cdots) \cdot \alpha_{k-\ell} \quad . \end{aligned}$$

Thus, in yet another basis E_k, $1 \leq k \leq g$, the invariant vector fields on Jac C are just given by

(10.7)' $$E_k\wp = \tfrac{1}{4}G_k(\wp,\wp^{(1)},\cdots).$$

(Here $E_k\wp^{(n)}$ is given by the rule

$$E_k\wp^{(n)} = E_kD^n\wp = D^nE_k\wp = G_k^{(n)}(\wp,\wp^{(1)},\cdots).)$$

At this point, we have found the link to the famous Korteweg-de Vries equation. Namely, we have

$$E_2\wp = \wp^{(3)} + 12\wp\cdot\wp^{(1)}.$$

This means that if $\wp$ is restricted to a 2-plane in Jac C tangent to the vector fields E_2 and $E_1 = D = '$, it gives a solution of the KdV equation

$$\frac{\partial}{\partial t}f(x,t) = \frac{\partial^3}{\partial x^3}f(x,t) + 12f(x,t)\cdot\frac{\partial}{\partial x}f(x,t).$$

We want to explore this link further in the last section.

§11. The Korteweg-deVries dynamical system.

As with the Neumann dynamical system, our purpose now is to introduce a dynamical system interesting in its own right, and then to show that it can, in some cases, be integrated explicitly by the theory of hyperelliptic Jacobians. More precisely, we can, following the ideas in the previous section, define an embedding of Jac C in an infinite dimensional space:

$$(\text{Jac}-\Theta) \longrightarrow \left(\begin{array}{c}\text{vector space } R_1 \text{ of analytic functions} \\ f(x) \text{ defined in some neigh. of } 0 \in \mathbb{C}\end{array}\right)$$

$$\vec{z}_0 \longmapsto \wp(\vec{z}_0+x\vec{e}) = \sum_{n=0}^{\infty} \wp^{(n)}(\vec{z}_0)\cdot \frac{x^n}{n!}$$

On R_1, we consider a simple class of vector fields X: those which assign to f a tangent vector in $X_f \in T_{R_1,f} \cong R_1$ given by

$$X_f = P(f,\dot{f},\cdots,f^{(n)}),\ P \text{ a polynomial.}$$

Integrating this vector field means finding an analytic function $f(x,y)$ s.t.

$$\frac{\partial f}{\partial y} = P\left(f, \frac{\partial f}{\partial x}, \ldots, \frac{\partial^n f}{\partial x^n}\right).$$

By the Cauchy-Kowalevski Theorem, for all $f(x,0)$ analytic in $|x| < \varepsilon$, there exists $f(x,y)$ analytic in $|x|,|y| < \eta$ solving this. What we want to do is to set up a sequence $X_1,X_2,\cdots$ of such vector fields called the Kortweg-de Vries hierarchy which a) commute $[X_i,X_j] = 0$ — we must define this carefully — and b) are Hamiltonian in a certain formal sense, such that c) for all g, and for all hyperelliptic curves C of genus g:

$$\mathrm{Im}(\mathrm{Jac}-\Theta) = \left[\begin{array}{l}\text{orbit of all flows } X_n, \\ \text{i.e., all } X_n \text{ are tangent to image} \\ \text{and a codimension g subspace of} \\ \sum c_n X_n \text{ are even 0 on Image}\end{array}\right]$$

In fact, d) in some sense "fixing the value of these Hamiltonians" gives the orbits of the X_n's: we will merely state some results of this type without proof. Thus $\{X_n\}$ may be considered an infinite-dimensional completely integrable Hamiltonian system.

We first investigate what it means for 2 such vector fields to commute. Let

$$X_f = P(f,\dot{f},\cdots,f^{(n)})$$
$$Y_f = Q(f,\dot{f},\cdots,f^{(m)}).$$

Then, starting at a function f, the path through f obtained by integrating X_f is

$$f(x,t) = f+tP\left(f,\dot{f},\cdots,f^{(n)}\right) + \frac{t^2}{2}\sum_{k=0}^{n} \frac{\partial P}{\partial f^{(k)}}\left(f,\cdots\right)\cdot\left(\frac{d}{dx}\right)^k P(f,\cdots) + \cdots\cdots$$

(because the t-derivative of the RHS is

$$\left[P(f,\dot{f},\cdots) + t\sum \frac{\partial P}{\partial f^{(k)}}\cdot\left(\frac{d}{dx}\right)^k \cdot P(f,\dot{f},\cdots)+\cdots\right] \equiv P(f+tP,\dot{f}+t\frac{d}{dx}P,\cdots\cdots) \bmod t^2).$$

To go in 2 directions at once, one must be able to define the (s.t)-term unambiguously, i.e., the coefficient of s.t in

$$t\,P(f+sQ,\ \dot{f}+s\dot{Q},\cdots\cdots)$$

and

$$s\,Q(f+tP,\ \dot{f}+t\dot{P},\cdots)$$

must be equal. This means

$$\text{(11.1)}\qquad \sum \frac{\partial P}{\partial f^{(k)}}\cdot\left(\frac{d}{dx}\right)^k Q \;=\; \sum \frac{\partial Q}{\partial f^{(k)}}\cdot\left(\frac{d}{dx}\right)^k P\,.$$

Theorem 11.2: (11.1) holds if and only if for all $f \in R_1$, there is an analytic function $f(x,s,t)$ (for $|x|,|s|,|t|<\varepsilon$) such that

$$\frac{\partial f}{\partial t} = P(f,\dot{f},\cdots)$$

$$\frac{\partial f}{\partial s} = Q(f,\dot{f},\cdots).$$

Proof: The existence of f implies (11.1) by working out the meaning of the equality of mixed derivatives

$$\left.\frac{\partial P}{\partial s}\right|_{s=t=0} = \left.\frac{\partial^2 f}{\partial s\partial t}\right|_{s=t=0} = \left.\frac{\partial Q}{\partial t}\right|_{s=t=0}$$

Given (11.1), we define f as follows:

a) let $R_3 = \mathbb{C}[X_0,X_1,X_2,\cdots]$ be a polynomial ring

b) let $\overline{}: R_3 \longrightarrow R_1$ be the map

$$\overline{X}_1 = \left(\frac{d}{dx}\right)^i f$$

Thus R_3 is a differential ring if we let $\dot{X}_i = X_{i+1}$, and $\overline{}$ is the homomorphism of differential rings carrying X_0 to f.

c) Let $D: R_3 \longrightarrow R_3$ be the derivation such that

$$D(X_0) = P(X_0, X_1, \cdots, X_n)$$
$$D(\dot{a}) = D(a)^{\bullet}$$

d) Likewise, let $E: R_3 \longrightarrow R_3$ be the derivation such that

$$E(X_0) = Q(X_0, X_1, \cdots, X_m)$$
$$E(\dot{a}) = E(a)^{\bullet}.$$

e) Let

$$\Phi(s,t) = \sum_{i,j\geq 0} \frac{D^i E^j (X_0)}{i!j!} t^i s^j \in R_3[[s,t]].$$

Note that (11.1) means precisely that $[D,E] = 0$ as derivations of R_3, and we check:

$$\frac{\partial \Phi}{\partial t} = D\Phi, \quad \frac{\partial \Phi}{\partial s} = E s.$$

Moreover,

$$\frac{\partial}{\partial t}(P(\cdots, \Phi^{(k)}, \cdots)) = \sum_k \frac{\partial P}{\partial X_k} \cdot \frac{\partial}{\partial t} \Phi^{(k)}$$

$$= \sum_k \frac{\partial P}{\partial X_k} \cdot D\,\Phi^{(k)}$$

$$= D(P(\cdots, \Phi^{(k)}, \cdots)).$$

Now both $P(\cdots, \Phi^{(k)}, \cdots)$ and $\frac{\partial \Phi}{\partial t}$ satisfy the equations

$$\frac{\partial}{\partial t}\Psi = D\Psi, \quad \frac{\partial}{\partial s}\Psi = Es \quad \text{and} \quad \Psi\big|_{s=t=0} = P(\cdots, X_k, \cdots),$$

hence they are equal and

$$\frac{\partial \Phi}{\partial t} = P(\cdots, \Phi^{(k)}, \cdots).$$

Likewise,

$$\frac{\partial\Phi}{\partial s} = Q(\cdots,\Phi^{(k)},\cdots).$$

Therefore

$$f(x,s,t) = \sum\frac{\overline{D^i E^j(X_0)}}{i!j!}\,t^i s^j$$

satisfies

$$\frac{\partial f}{\partial t} = P(\cdots,(\frac{d}{dx})^k F,\cdots), \quad \frac{\partial f}{\partial s} = Q(\cdots,(\frac{d}{dx})^k F,\cdots).$$

QED

Thus the differential ring R_1 is very convenient for integrating flows. However, the Hamiltonians that define these flows do not exist on R_1. Instead, we need

$$R_2 = \left\{\begin{matrix}\text{(differential) ring of } C^\infty \text{ functions}\\ f(x) \text{ with compact support}\end{matrix}\right\}$$

R_2 has the advantage that there is a large class of functions (usually called "functionals")

$$\phi_P:\ R_2 \longrightarrow \mathbb{C},$$

namely

$$\phi_P(f) = \int_{-\infty}^{+\infty} P(x,f,\dot{f},\cdots,f^{(n)})\,dx$$

where P is a polynomial in $f,\cdots,f^{(n)}$, whose coefficients are C^∞ functions of x, and whose constant term has compact support. These functionals may be called C^∞-functionals because by the calculus of variations they have excellent derivatives: i.e.,

$$\lim_{\varepsilon\to 0} \frac{\phi_P(f+\varepsilon g)-\phi_P(f)}{\varepsilon} = \int_{-\infty}^{+\infty} \sum_{k=0}^{n} \frac{\partial P}{\partial f^{(k)}} \cdot g^{(k)} dx$$

$$= \int_{-\infty}^{+\infty}\left(\sum_{k=0}^{n} (-1)^k \left(\frac{d}{dx}\right)^k \left(\frac{\partial P}{\partial f^{(k)}}\right)\right)\cdot g(x)\, dx$$

(integration by parts).

Define

$$\frac{\delta P}{\delta f} = \sum_{k=0}^{n} (-1)^k \left(\frac{d}{dx}\right)^k \left(\frac{\partial P}{\partial f^{(k)}}\right)$$

to be the variational derivative of P. We want to set up a co-symplectic structure on R_2, and define vector fields V_{ϕ_P} for these Hamiltonians. These will turn out to be examples of the same type of vector fields that we considered on R_1: but on R_1, they can be integrated locally, on R_2 they come from Hamiltonians. At the very end of this §, we will mention briefly yet another approach: that of McKean, Van Moerbecke and Trubowitz, who used R_4 = ring of periodic C^∞-functions f(x), and could do both at once.

However, the clearest and most elegant way to bring in the co-symplectic structure is in a much larger vector space: a space of differential operators. This approach goes back to Lax and Gel'fand-Dikii and has been highly developed by M. Adler and Lebedev-Manin*. Up to a point, we can develop the theory for any of our differential rings R, but later we will restrict to R_2.

---

* Mark Adler, On a Trace Functional for Formal Pseudo-Differential Operators and the Symplectic Structure of the Korteweg-DeVries Type Equations, Inv. Math., 50, (1979), p. 219;
J.I. Manin, Algebraic Aspects of non-linear differential equations, Modern Problems in Mathematics, (VINITI, 1978)

The central idea is to associate to $f \in R$ the differential operator

$$\left(\frac{d}{dx}\right)^2 + f(x)$$

and to consider R as part of an even bigger space, viz.

$$R[D] = \left[\begin{array}{l}\text{vector space of all differential operators} \\ \sum_{n=0}^{d} a_n(x) D^n, \quad D = \frac{d}{dx} \\ a_n(x) \in R\end{array}\right]$$

In fact, we put this in a yet bigger space:

$$R\{D\} = \left[\begin{array}{l}\text{vector space of "pseudo-differential} \\ \text{operators symbols"} \\ \sum_{n=-\infty}^{d} a_n(x) D^n, \quad a_n(x) \in R\end{array}\right]$$

In R{D}, we can introduce a ring structure as follows:

Note that

$$D(fg) = fDg + gDf.$$

Thus as operators on R,

(11.3) $$\underline{D \circ f = f \circ D + \dot{f}} \ .$$

Taking this as our golden rule, we get a ring structure on R[D] such that:

$$\begin{aligned}(fD^n)\circ(gD^m) &= \sum_{k=0}^{n} \binom{n}{k} f.g^{(k)}.D^{n+m-k} \\ &= \sum_{k=0}^{n} \frac{1}{k!} \partial_D^k(fD^n) * \partial_x^k(gD^m)\end{aligned}$$

or more generally

$$(11.4)\qquad X\circ Y = \sum_{k=0}^{\infty} \frac{1}{k!}\, \partial_D^k X * \partial_x^k Y$$

$*$ = multiply by as though f,D commute.

In fact, this extends to $R\{D\}$ too, if we extend (11.1) via

$$f\circ D^{-1} = D^{-1}\circ f + D^{-1}\circ \dot{f}\circ D^{-1}$$

hence

$$\begin{aligned} D^{-1}\circ f &= f\circ D^{-1} - D^{-1}\circ\dot{f}\circ D^{-1} \\ &= f\circ D^{-1} - \dot{f}\circ D^{-2} + D^{-1}\circ\ddot{f}\circ D^{-2} \\ &= f\circ D^{-1} - \dot{f}\circ D^{-2} + \ddot{f}\circ D^{-3} - D^{-1}\circ\dddot{f}\circ D^{-3}\ , \\ &\cdots\cdots \end{aligned}$$

i.e.,

$$(11.3)'\qquad D^{-1}\circ f = [f\circ D^{-1} - \dot{f}\circ D^{-2} + \ddot{f}\circ D^{-3} + \cdots + (-1)^k f^{(k)}\circ D^{-k-1} + \cdots]$$

Note that again

$$D^{-1} f = \sum_{k=0}^{\infty} \frac{1}{k!}\, \partial_D^k (D^{-1}) * \partial_x^k (f)$$

so the general rule for mult. is still (11.2):

$$X\circ Y = \sum_{k=0}^{\infty} \frac{1}{k!}\, \partial_D^k X * \partial_x^k Y\quad .$$

Associativity is very easy to check:

$$\begin{aligned} (X\circ Y)\circ Z &= \sum \frac{1}{\ell!}\, \partial_D^\ell \Big(\sum \frac{1}{k!}\, \partial_D^k X * \partial_x^k Y\Big) * \partial_x^\ell Z \\ &= \sum \frac{1}{k!\,p!\,(\ell-p)!}\, \partial_D^{k+p} X * \partial_x^k \partial_D^{\ell-p} Y * \partial_x^\ell Z \end{aligned}$$

$$\begin{aligned} X\circ(Y\circ Z) &= \sum \frac{1}{k!}\, \partial_D^k X * \partial_x^k \Big(\sum \frac{1}{\ell!}\, \partial_D^\ell Y * \partial_x^\ell Z\Big) \\ &= \sum \frac{1}{\ell!\,p!\,(k-p)!}\, \partial_D^k X * \partial_D^\ell \partial_x^{k-p} Y * \partial_x^{p+\ell} Z \\ &= \sum \frac{1}{(\ell-p)!\,p!\,k!}\, \partial_D^{k+p} X * \partial_D^{\ell-p} \partial_x^k Y * \partial_x^\ell Z \end{aligned}$$

Proposition 11.5: For all $d \in \mathbb{Z}$, every element

$$X = D^d + a_1 D^{d-1} + \cdots \in R\{D\}$$

has an inverse

$$X^{-1} = D^{-d} + b_1 D^{-d-1} + \cdots\cdots \in R\{D\}.$$

Corollary 11.6. The set of elements

$$1 + a_1 D^{-1} + a_2 D^{-1} + \cdots\cdots\cdots$$

in $R\{D\}$ is a group $\mathcal{G}$, called the Volterra group by Lebedev-Manin. Lie $\mathcal{G} \underset{\text{def}}{=} \{a_1 D^{-1} + a_2 D^{-2} + \cdots\}$ is a Lie algebra under [,].

Proof of Prop.: Construct D^{-1} by induction. Suppose we have $b_1, \cdots, b_n$ such that

$$(D^{-d} + b_1 D^{-d-1} + \cdots + b_n D^{-d-n}) \circ (D^d + a_1 D^{d-1} + \cdots\cdot) = 1 + cD^{-n-1} + \cdots\cdot.$$

Then it follows that

$$(D^{-d} + b_1 D^{-d-1} + \cdots + b_n D^{-d-n} - cD^{-d-n-1}) \circ (D^d + a_1 D^{d-1} + \cdots) = 1 + (\text{terms in } D^{-n-2} \text{ or lower}).$$

QED

For instance, one checks that

$$(D^2 + q)^{-1} = D^{-2} - qD^{-4} + 2\dot{q}D^{-5} + \cdots\cdots .$$

The following Proposition is due*, in fact, to I. Schur in 1904, as P.M. Cohn pointed out to me:

---

* I. Schur, Über vertauschbare lineare Differentialausdrücke, Berliner Math. Ges. Sitzber. 3 (Archiv der Math. Beilage (3) 8) (1904), pp. 2-8.

Proposition 11.7: For all $d \geq 1$ and all

$$X = D^d + a_1 D^{d-1} + \cdots\cdots \in R\{(D)\}$$

X has a unique $d^{\underline{th}}$ root

$$X^{1/d} = D + b_1 + b_2 D^{-1} + \cdots\cdot \in R\{D\}$$

and the commutator $Z(X)$ of X in $R\{D\}$ is the ring of Laurent series

$$\sum_{i=-\infty}^{n} c_i X^{i/d}, \qquad c_i \in \mathbb{C}.$$

Proof: The main point is the calculation:

$$[X, cD^m] = d\,\dot{c}\,D^{d+m-1} + \text{lower terms},\ c \in R.$$

From this it follows by easy induction that $Z(X)$ has, mod scalars and lower order terms, a unique element of each degree $m \in \mathbb{Z}$ and that it has the form $(cD^m + \text{lower terms})$ $c \in \mathbb{C}$. If $Y \in Z(X)$ has degree 1, $Y' \in Z(X)$ has degree -1, it follows that $Y \circ Y' = c+W$, $c \in \mathbb{C}$, $c \neq 0$ and $\deg W < 0$. Therefore

$$Y^{-1} = \frac{1}{c} Y \cdot \sum_{i=0}^{\infty} (-1)^i W^i / c^i \in Z(X),$$

hence

$$Z(X) \supset \{\text{ring of Laurent series } \sum_{i=-\infty}^{n} c_i Y^i\}$$

hence "=" holds here because each side has one new element in each degree. Thus $Z(X)$ is commutative. Finally, X itself is in $Z(X)$ so

$$X = \sum_{i=-\infty}^{d} c_i Y^i, \quad c_i \in \mathbb{C},\ c_d \neq 0$$

and, in a ring of Laurent series, such an element has a unique $d^{\underline{th}}$ root (up to root of unity):

$$X^{1/d} = c_d^{1/d} Y \circ \left(1 + \frac{c_{d-1}}{c_d} Y^{-1} + \frac{c_{d-2}}{c_d} Y^{-2} + \cdots\right)^{1/d}$$

where the last term can be expanded by the binomial theorem. QED

Returning to the $2^{\underline{nd}}$ order operator $X = D^2+q$, we can calculate $\sqrt{X}$ in terms of the universal polynomials introduced in §10. In fact, expand:

$$\sqrt{D^2+q} = \sum_{n=0}^{\infty} \left(f_n(q,\dot{q},\cdots)D - \frac{g_n(q,\dot{q},\cdots)}{2}\right)(D^2+q)^{-n}$$

where f_n, g_n are universal polynomials without constant term, except for $f_0 = 1$. (Also $g_o = 0$). Now

$$\begin{aligned}
\sum_0^{\infty} \left(f_n D - \frac{g_n}{2}\right) \circ (D^2+q)^{-n+1} &= \sqrt{D^2+q} \circ (D^2+q) \\
&= (D^2+q) \circ \sqrt{D^2+q} \\
&= \sum (D^2+q) \circ \left(f_n D - \frac{g_n}{2}\right) \circ (D^2+q)^{-n} \\
&= \sum_0^{\infty} \left(f_n D - \frac{g_n}{2}\right) \circ (D^2+q)^{-n+1} \\
&\quad + \sum_{n=0}^{\infty} \left(\ddot{f}_n D + 2\dot{f}_n D^2 - f_n \dot{q} - \frac{\ddot{g}_n}{2} - \dot{g}_n D\right) \circ (D^2+q)^{-n}
\end{aligned}$$

hence

$$0 = \sum \left((\ddot{f}_n - \dot{g}_n)D + 2\dot{f}_n(D^2+q) - 2\dot{f}_n q - f_n \dot{q} - \frac{\ddot{g}_n}{2}\right) \circ (D^2+q)^{-n}$$

hence

(11.8)
$$g_n = \dot f_n$$
$$\dot f_{n+1} = \dot f_n q + \tfrac{1}{2} f_n \dot q + \frac{\ddot g_n}{4} .$$

Thus, relating this to §10, if $q = 2\wp$, then

$$4^n f_n(q,\dot q,\cdots) = F_n(\tfrac{q}{2}, \tfrac{\dot q}{2},\cdots\cdot)$$
$$4^n g_n(q,\dot q,\cdots) = G_n(\tfrac{q}{2}, \tfrac{\dot q}{2},\cdots\cdot) .$$

One may while away an hour or more calculating this out a ways:

$$\sqrt{D^2+q} = D + (\tfrac{q}{2}D - \tfrac{\dot q}{4})\circ(D^2+q)^{-1} + (\tfrac{3q^2+\ddot q}{8}D - \tfrac{6q\dot q+\dddot q}{16})\circ(D^2+q)^{-2} +\cdots$$
$$= D + \tfrac{q}{2}D^{-1} - \tfrac{\dot q}{4}D^{-2} + (\tfrac{\ddot q-q^2}{8})D^{-3} + (\tfrac{6q\dot q-\dddot q}{16})D^{-4} +\cdots\cdots .$$

We now choose our R to be the ring R_2 of C^∞-functions on $\mathbb{R}$ with compact support. This enables us to integrate elements of R as well as differentiate them. Many of our conclusions will however be quite formal and for these we may afterwards go back to the original R.

In calculations in $R_2\{D\}$, we find that the coefficient of D^{-1} has a very important special property, viz.:

Theorem 11.9 (Adler): For all $X,Y \in R_2\{D\}$, the coefficient a_{-1} of D^{-1} in X,Y is the derivative of a polynomial in the coefficients of X and Y, hence

$$\int_{-\infty}^{+\infty} a_{-1}(x)\,dx = 0.$$

<u>Proof</u>: By linearity, it suffices to consider the case

$$X = aD^k, \quad Y = bD^\ell.$$

Clearly, if $k+\ell \leq -1$ or if $k \geq 0, \ell \geq 0$ there is nothing to check. We may as well suppose $k \geq 1, \ell \leq -1$ and use:

$$X \circ Y = \sum_{n=0}^{\infty} \binom{k}{n} ab^{(n)} D^{k+\ell-n}, \qquad \text{with term } k(k-1)\cdots(-\ell)ab^{(k+\ell+1)}D^{-1}.$$

Likewise, $\ell(\ell-1)\cdots\cdot(-k)ba^{(k+\ell+1)}$ is the coefficient of D^{-1} in $Y \circ X$. The difference is $ab^{(k+\ell+1)} + (-1)^{k+\ell}a^{(k+\ell+1)}\cdot b$ which is the derivative of $[ab^{(k+\ell)} - \dot{a}\, b^{(k+\ell-1)} + \ddot{a}\, b^{(k+\ell-2)} - \cdots + (-1)^{k+\ell}a^{(k+\ell)}\cdot b]$.

<u>QED</u>

We define

$$\mathrm{tr}: \; R_2\{D\} \longrightarrow \mathbb{C}$$

by

$$\mathrm{tr}\, X = \int_{-\infty}^{+\infty} a_1(x)\,dx, \qquad \text{if } X = \sum a_k D^k .$$

Now put the vector spaces

$$R_2[D], \qquad \mathrm{Lie}\, \mathfrak{g}$$

in duality by

$$\langle X, Y \rangle = \mathrm{tr}(X \circ Y) .$$

In particular, if

$$X = \sum_{n=0}^{d} a_n D^n, \qquad Y = \sum_{n=0}^{\infty} D^{-n-1} b_n$$

then

$$\langle X,Y\rangle = \int \sum_{n=0}^{d} (a_n b_n)\,dx$$

so that $R_2[D]$ is isomorphic to a subspace of $(\text{Lie}\,\mathfrak{g})^*$, the linear functions ℓ on Lie $\mathfrak{g}$. Thus the lie algebra Lie $\mathfrak{g}$ acts on Lie $\mathfrak{g}$ by the adjoint representation $ad_X(Y) = [X,Y]$ and on $R_2[D]$ by the co-adjoint representation. Explicitly, for all $Y \in$ Lie $\mathfrak{g}$, define

$$ad_Y^*: R_2[D] \longrightarrow R_2[D]$$

by

$$\langle ad_{Y_1}^*(X), Y_2\rangle = -\langle X, ad_{Y_1}(Y_2)\rangle$$
$$= -\langle X, [Y_1, Y_2]\rangle.$$

Let

$$+: \{R_2\}D \longrightarrow R_2\ D$$

be the projection $\left(\sum_{n=-\infty}^{d} a_n D^n\right)_+ = \sum_{n=0}^{d} a_n D^n.$

<u>Corollary</u> 11.10: $ad_Y^*(X) = [Y,X]_+$.

<u>Proof</u>: $\langle ad_{Y_1}^*(X), Y_2\rangle = -\langle X, [Y_1, Y_2]\rangle$

$$= -\text{tr}(X \circ Y_1 \circ Y_2 - X \circ Y_2 \circ Y_1)$$
$$= -\text{tr}((X \circ Y_1 - Y_1 \circ X) \circ Y_2)$$
$$= -\text{tr}([X, Y_1]_+ \circ Y_2) \quad \text{since deg } \tfrac{1}{2} \leq -1$$
$$= \langle [Y_1, X]_+, Y_2\rangle.$$

<u>QED</u>

We now recall a very general construction due to Kostant and Kirillov which has many important applications. Let G be an ordinary Lie group and $\mathfrak{g}$ its Lie algebra (which is finite dimensional). Then $\mathfrak{g}^*$ has a "co-symplectic structure" on it. We explain this quite carefully to facilitate the infinite-dimensional version to be used below:

a) $\forall x \in \mathfrak{g}^*$, identify $T_{\mathfrak{g}^*,x} \cong \mathfrak{g}^*$,

$$\text{hence } T^*_{\mathfrak{g}^*,x} \cong \mathfrak{g} .$$

Then for all $\alpha,\beta \in T^*_{\mathfrak{g}^*,x} \cong \mathfrak{g}$, define

$$\Omega^*_x(\alpha,\beta) = \langle x,[\alpha,\beta]\rangle .$$

Thus Ω^*_x is a skew-symmetric bilinear form on $T^*_{\mathfrak{g}^*,x}$.

b) Now for all functions f,g on $\mathfrak{g}^*$, we get $df_x, dg_x \in T^*_{\mathfrak{g}^*,x} \cong \mathfrak{g}$, namely

$$\langle y, df_x \rangle = \lim \frac{f(x+\varepsilon y)-f(x)}{\varepsilon}$$

$$\langle y, dg_x \rangle = \lim \frac{g(x+\varepsilon y)-g(x)}{\varepsilon} .$$

Hence we define the Poisson bracket:

$$\{f,g\}_x = \Omega^*_x(df_x,dg_x)$$

$$= \langle x,[df_x,dg_x]\rangle .$$

c) For all functions f on $\mathfrak{g}^*$, this gives a vector field V_f on $\mathfrak{g}^*$. Namely, if $x \in \mathfrak{g}^*$, then $(V_f)_x \in T_{\mathfrak{g}^*,x} \cong \mathfrak{g}^*$ is given by

$$\begin{aligned} \langle (V_f)_x, \beta\rangle &= \Omega^*_x(df_x, \beta) \\ &= \langle x, [df_x, \beta]\rangle \\ &= \langle x, \mathrm{ad}_{df_x}(\beta)\rangle \\ &= -\langle \mathrm{ad}^*_{df_x}(x), \beta\rangle \end{aligned}$$

or

$$(V_f)_x = -\mathrm{ad}^*_{df_x}(x).$$

Note that for all functions g on $\mathfrak{g}^*$,

$$\begin{aligned} V_f^*(g) &\underset{\mathrm{def}}{=} \langle V_f, dg\rangle \\ &= \Omega^*(df, dg) \\ &= \{f,g\}, \end{aligned}$$

hence

$$\{f,g\}_{x_0} = \frac{d}{d\varepsilon}\, g\big(x_0+\varepsilon (V_f)_{x_0}\big)\Big|_{\varepsilon=0}.$$

d) Moreover, given any 2 vector fields V_1, V_2, we get their bracket $[V_1, V_2] = V_3$, which may be defined equivalently as

$$V_3^*(f) = V_1(V_2(f)) - V_2(V_1(f)), \text{ all functions f on } \mathfrak{g}^*$$

or directly by:

$$V_{3,x_0} = \lim \frac{V_{2,x_0+\varepsilon x_1} - V_{2,x_0}}{\varepsilon} - \lim \frac{V_{1,x_0+\varepsilon x_2} - V_{1,x_0}}{\varepsilon}$$

where $x_1 = V_{1,x_0}$, $x_2 = V_{2,x_0}$.

The basic result -that this is a "good" co-symplectic structure- is that

(11.11) $$\{f,\{g,h\}\}+\{g,\{h,f\}\}+\{h,\{f,g\}\}=0$$

or equivalently

$$[V_f,V_g]=V_{\{f,g\}} \quad .$$

We prove this in 2 steps:

Step I: For all $\alpha \in \mathfrak{g}$, let ℓ_α be the linear fcn. on $\mathfrak{g}^*$ given by $\ell_\alpha(x) = \langle x,\alpha\rangle$. Then $(d\ell_\alpha)_x = \alpha$ for all $x \in \mathfrak{g}^*$, and the definition tells us immediately

$$\{\ell_\alpha,\ell_\beta\} = \ell_{[\alpha,\beta]}.$$

Thus, Jacobi's identity on $\mathfrak{g}$ gives us (11.9) when f,g,h are ℓ_α's.

Step II: We prove (11.9) at a point x_0 under the assumption that $(df)_{x_0} = 0$. It merely states the equality of the mixed $2^{\underline{nd}}$ derivatives of f: i.e., let $(d^2f)_x$ be the $2^{\underline{nd}}$ derivative:

$$(d^2f)_x(y,z) = \frac{\partial^2}{\partial\varepsilon\partial\eta} f(x+\varepsilon y+\eta z)\Big|_{(0,0)} \quad .$$

Then

$$\{g,\{h,f\}\} = \frac{\partial}{\partial\varepsilon}(\{h,f\}(x_0+\varepsilon(X_g)_{x_0}))\Big|_{\varepsilon=0}$$

$$= \frac{\partial^2}{\partial\varepsilon\partial\eta} f\Big(x_0+\varepsilon(X_g)_{x_0} + \eta(X_h)_{x_0+\varepsilon(X_g)_{x_0}}\Big)\Big|_{\varepsilon=\eta=0}$$

$$= \frac{\partial^2}{\partial\varepsilon\partial\eta} f\Big(x_0+\varepsilon(X_g)_{x_0}+\eta(X_h)_{x_0} + \underbrace{\varepsilon\eta\,(\text{pt. of } \mathfrak{g}^* \text{ depending in } C^\infty \text{ way on } \varepsilon)}_{\text{ignore this because } df_x = 0}\Big)\Big|_{\varepsilon=\eta=0}$$

$$= (d^2f)_{x_0}((X_g)_{x_0},(X_h)_{x_0}).$$

Thus

$$\{g,\{h,f\}\} = \{h,\{g,f\}\} \qquad \text{and} \quad df_{x_0} = 0 \Longrightarrow \{f,\{g,h\}\} = 0.$$

QED

Rather surprisingly, all of this works without essential change for the infinite-dimensional case.

$$\mathfrak{g}^* = R_2[D]$$
$$\mathfrak{g} = \text{Lie}\ \mathcal{G} \ ,$$

provided we restrict ourselves to an appropriate class of functions on $R_2[D]$. We use the maps

$$\phi_P: \ R_2[D] \longrightarrow \mathbb{C}$$

$$\phi_P(X) = \int_{-\infty}^{+\infty} P(x,\cdots,a_k^{(\ell)},\cdots)dx, \quad \text{if } X = \sum_{k=0}^{d} a_k D^k$$

P a C^∞-function depending on a finite number of the $a_k^{(\ell)}$'s

The main point is that, as above, ϕ_P is sufficiently differentiable:

$$\phi_P(X+\varepsilon Y) = \int P(x,\cdots,(a_k+\varepsilon b_k)^{(\ell)},\cdots)dx$$

is a C^∞ function of ε and:

$$\frac{d}{d\varepsilon}\phi_P(X+\varepsilon Y) = \int \sum_{k,\ell} \frac{\partial P}{\partial a_k^{(\ell)}} \cdot b^{(\ell)} dx$$

$$= \int \sum_k \left(\sum_\ell (-1)^\ell \left(\frac{d}{dx}\right)^\ell \frac{\partial P}{\partial a_k^{(\ell)}}\right) \cdot b_k \, dx$$

$$= \left\langle Y, \sum D^{-k} \cdot \left(\sum_\ell (-1)^\ell \left(\frac{d}{dx}\right)^\ell \frac{\partial P}{\partial a_k^{(\ell)}}\right)\right\rangle,$$

hence

$$(d\phi_P)_X = \sum_k D^{-k}\left(\sum_\ell (-1)^\ell \left(\frac{d}{dx}\right)^\ell \frac{\partial P}{\partial a_k^{(\ell)}}\right) \in \operatorname{Lie} \mathfrak{g} .$$

By (11.10), the corresponding vector field V_P is just:

$$(V_{\phi_P})_X = \left[X, \sum D^{-k}\left(\sum_\ell (-1)^\ell \left(\frac{d}{dx}\right)^\ell \frac{\partial P}{\partial a_k^{(\ell)}}\right)\right]_+$$

$$\{\phi_P,\phi_Q\}_X = \int\left(X \circ \left[\sum_{k,\ell} D^{-k}(-1)^\ell\left(\frac{d}{dx}\right)^\ell \frac{\partial P}{\partial a_k^{(\ell)}}, \sum_{k,\ell} D^{-k}(-1)^\ell\left(\frac{d}{dx}\right)^\ell \frac{\partial Q}{\partial q^{(\ell)}}\right]\right)_{-1} dx.$$

The Jacobi identity for { , } and the formula

$$[V_{\phi_P}, V_{\phi_Q}] = V_{\{\phi_P,\phi_Q\}}$$

are proven exactly as before.

We now specialize all this to the submanifold M of R[D]:

$$M = \{D^2+q \mid q \in R\} .$$

In general, one cannot restrict a co-symplectic structure from a space N to a submanifold M unless for all $x \in M$, the 2-form Ω_x^* factors through $T^*_{M,x}$:

$$\begin{array}{ccc} \Omega_x^*: \ T^*_{N,x} \times T^*_{N,x} & \longrightarrow & \mathbb{C} \\ \searrow & & \nearrow \\ & T^*_{M,x}\times T^*_{M,x} & \end{array} .$$

But if we compute $\Omega^*_{D^2+q}$ we find the following. For all $\alpha,\beta \in \operatorname{Lie}\mathfrak{g}$

$$\Omega^*_{D^2+q}(\alpha,\beta) = \langle D^2+q, [\alpha,\beta]\rangle .$$

Let $\alpha = D^{-1}\alpha_0 + D^{-2}\alpha_1 + \cdots$

$\beta = D^{-1}\beta_0 + D^{-2}\beta_1 + \cdots$

then
$$\begin{aligned}[\alpha,\beta] &= D^{-1}\alpha_0 D^{-1}\beta_0 - D^{-1}\beta_0 D^{-1}\alpha_0 + \cdots \\ &= (D^{-2}\alpha_0\beta_0 + D^{-3}\dot{\alpha}_0\beta_0 + \cdots) - (D^{-2}\alpha_0\beta_0 + D^{-3}\dot{\beta}_0\alpha_0 + \cdots) \\ &= D^{-3}\cdot(\dot{\alpha}_0\beta_0 - \dot{\beta}_0\alpha_0) + \text{higher terms,}\end{aligned}$$

so

$$\begin{aligned}\Omega^*_{D^2+q}(\alpha,\beta) &= \int (\dot{\alpha}_0\beta_0 - \dot{\beta}_0\alpha_0)\,dx \\ &= 2\int \dot{\alpha}_0\beta_0\,dx.\end{aligned}$$

This depends only on α_0, β_0, which give the restriction to α,β to linear functions on T_{M,D^2+q}. Thus we have a co-symplectic structure on M. In fact, it is non-degenerate now. This non-degenerate 2-form was discovered by Gardner and Greene.

Now for all functionals on M:

$$\phi_P(D^2+q) = \int P(x,q,\dot{q},\cdots,q^{(n)})\,dx$$

we see that using the variational derivative $\delta P/\delta q$ defined above:

$$(11.12)\qquad \left|\begin{aligned} (d\phi_P)_{D+q} &= D^{-1}\cdot\frac{\delta P}{\delta q} \\ \{\phi_P,\phi_Q\}_{D^2+q} &= 2\int \left(\frac{\delta \dot{P}}{\delta q}\right)\cdot\left(\frac{\delta Q}{\delta q}\right)dx \\ (V_{\phi_P})_{D+q} &= \left[D^2+q,\ D^{-1}\frac{\delta P}{\delta q}\right]_+ \\ &= D\frac{\delta P}{\delta q} - D^{-1}\frac{\delta P}{\delta q}D^2 \\ &= 2\left(\frac{\delta P}{\delta q}\right)^{\bullet}. \end{aligned}\right.$$

We will also have occasion to compute the bracket of 2 vector fields on M directly. Suppose $P_1(x,q,\dot{q},\cdots,q^{(n)})$ and $P_2(x,q,\dot{q},\cdots,q^{(n)})$ define 2 vector fields V_1,V_2 by the rule

$$(V_1)_{D^2+q} = P_1(x,q,\dot{q},\cdots,q^{(n)})$$

$$(V_2)_{D^2+q} = P_2(x,q,\dot{q},\cdots,q^{(n)})$$

then we can compute $[V_1,V_2]$ directly as in (d) above:

$$[V_1,V_2]_{D^2+q} = \lim \frac{P_2(x,q+\varepsilon P(x,q))-P_2(x,q)}{\varepsilon} - \lim \frac{P_1(x,q+\varepsilon P_2(x,q))-P_1(x,q))}{\varepsilon}$$

(11.12)'

$$= \sum_{k=0}^{\infty}\left[\frac{\partial P_2}{\partial q^{(k)}}\cdot P_1^{(k)} - \frac{\partial P_1}{\partial q^{(k)}}\cdot P_2^{(k)}\right]$$

which is a vector field of the same form.

Formulae (11.12) have the following consequence:

<u>Proposition</u> 11.13. <u>Given</u> $P(x,q,\dot{q},\cdots,q^{(n)})$ <u>and</u> $Q(x,q,\dot{q},\cdots,q^{(m)})$, <u>if there exists a polynomial</u> $H(x,q,\dot{q},\cdots,q^{(k)})$ <u>such that</u>

$$\left(\frac{\delta P}{\delta q}\right)^{\bullet}\cdot\left(\frac{\partial Q}{\partial q}\right) = \dot{H},$$

<u>then</u> $\{\phi_P,\phi_Q\} = 0.$

In fact, if P and Q are polynomials in the $q^{(k)}$ alone and don't involve x, the converse is true. This can be proven as follows. We use a purely formal result of the variational calculus in the differential ring

$$R_3 = \mathbb{C}[X_0, X_1, X_2, \cdots], \quad \dot{X}_i = X_{i+1}.$$

<u>Theorem</u> 11.14. <u>The sequence</u>

$$R_3 \xrightarrow{\;\cdot\;} R_3 \xrightarrow{\;\delta/\delta X\;} R_3$$

<u>is exact</u>, i.e., <u>for all polynomials</u> $f(X_0, X_1, \cdots, X_n)$,

$$\frac{\delta f}{\delta X} = 0 \iff f = \dot{g}, \text{ \underline{some} } g.$$

<u>Sketch of proof</u>: Working over R_2, we see that

$$f = \dot{g} \Longrightarrow \phi_f \equiv 0 \Longrightarrow \text{derivative } \frac{\delta f}{\delta q} \text{ of } \phi_f \text{ is zero.}$$

Since this is purely formal, it holds in R_3 too. To prove the converse, use induction on the order n of the highest derivatives in f, and argue like this:

$$\frac{\delta f}{\delta x} = 0 \Longrightarrow f = X_n \cdot f_1 + f_3 \Longrightarrow \exists g \text{ s.t. } f - \dot{g} \in \mathbb{C}[X_0, \cdots, X_{n-1}]$$
$$f_1, f_2 \in \mathbb{C}[X_0, \cdots, X_{n-1}]$$

<u>QED</u>

<u>Corollary</u> 11.15. <u>If</u> $P(q, \dot{q}, \cdots, q^{(n)})$, $Q(q, \dot{q}, \cdots, q^{(n)})$ <u>are polynomials, then</u>

$$\{\phi_P, \phi_Q\} = 0 \Longrightarrow \exists \text{ \underline{polynomial} } H(q, \dot{q}, \cdots, q^{(k)}) \text{ \underline{such that}}$$

$$\left(\frac{\delta P}{\delta q}\right)^{\bullet} \cdot \left(\frac{\delta Q}{\delta q}\right) = \dot{H}.$$

<u>Proof</u>: $\{\phi_P, \phi_Q\} = 0$ implies $d\{\phi_P, \phi_Q\} = 0$, i.e.,

$$\frac{\delta}{\delta q}\left(\left(\frac{\delta P}{\delta q}\right)^{\bullet} \cdot \left(\frac{\delta Q}{\delta q}\right)\right) = 0,$$

hence H exists by the Theorem. <u>QED</u>

This completes our general discussion of the co-symplectic structure on M. We now introduce the Korteweg-de Vries Hamiltonians:

$$H_n = \left(\frac{1}{n+1/2}\right) \operatorname{tr}\left((D^2+q)^{n+1/2}\right) . \tag{11.16}$$

Expanding $\sqrt{D^2+q}$ as above, we see that

$$H_n = \frac{1}{n+\frac{1}{2}} \operatorname{tr}\left((D^2+q)^{\frac{1}{2}} \circ (D^2+q)^n\right)$$

$$= \frac{1}{n+\frac{1}{2}} \sum_{k=0}^{\infty} \operatorname{tr}\left(f_k D - \frac{g_k}{2}\right) \circ (D^2+q)^{n-k}.$$

Note that $n-k \geq 0$, the kth term is a differential operator, hence has no trace and if $n-k \leq 2$, the kth term involves D^{-3} and lower, so has no trace. Therefore

$$H_n = \frac{1}{n+\frac{1}{2}} \operatorname{tr}\left(\left(f_{n+1}D - \frac{g_{n+1}}{2}\right) \circ (D^2+q)^{-1}\right)$$

$$= \frac{1}{n+\frac{1}{2}} \operatorname{tr}(f_{n+1}D^{-1})$$

$$= \frac{1}{n+\frac{1}{2}} \int f_{n+1}(q,\dot{q},\cdots)\,dx$$

$$= \phi_{(f_{n+1}/n+\frac{1}{2})}$$

which is a function on M of the type we are considering. We want to calculate the derivative of H_n:

<u>Lemma</u> 11.17.
$$\frac{\delta}{\delta q}\left[(D^2+q)^{n+\frac{1}{2}}\right]_{-1} = (n+\tfrac{1}{2})\left[(D^2+q)^{n-\frac{1}{2}}\right]_{-1} ,$$

i.e., $$\frac{\delta f_{n+1}}{\delta q} = (n+\tfrac{1}{2}) f_n$$

or $$(dH_n)_{D+q} = D^{-1} \circ f_n(q,\dot{q},\cdots).$$

Proof: This amounts to saying that for all $a(x) \in R$,

$$\frac{d}{d\varepsilon}\, tr\left((D^2+q+\varepsilon a)^{n+\frac12}\right) = (n+\tfrac12)\, tr\left((D^2+q)^{n-\frac12}\circ a\right).$$

Write

$$\left(D^2+q+\varepsilon a\right)^{\frac12} = E+\varepsilon E_1 \pmod{\varepsilon^2}.$$

Then

$$\begin{aligned}\frac{d}{d\varepsilon}\, tr(E^n \circ (E+\varepsilon E_1)^m) &= \sum tr\Big(E^n\circ \overbrace{E\circ\cdots\circ E\circ E_1\circ E\circ\cdots\circ E}^{m}\Big)\\ &= m\; tr\left(E^{n+m-1}\circ E_1\right)\;;\end{aligned}$$

especially for $m = 2$, this says

$$tr(E^n\circ a) = \frac{1}{d\varepsilon}\, tr(E^n\circ(D^2+q+\varepsilon a)) = 2\; tr(E^{n+1}\circ E_1)$$

so

$$\frac{d}{d\varepsilon}\; tr(E^n\circ (E+\varepsilon E_1)^m) = \frac{m}{2}\; tr(E^{n+m-2}\circ a).$$

In particular, if $n = 0$, we get

$$\frac{d}{d\varepsilon}\; tr\left((D^2+q+\varepsilon a)^{m/2}\right) = \frac{m}{2}\; tr\left((D^2+q)^{\frac{m}{2}-1}\circ a\right). \qquad \text{QED}$$

Theorem 11.18. a) $(V_{H_n})_{D+q} = 2g_n(q,\dot q,\cdots) = -\left[(D^2+q),\ [(D^2+q)^{n-\frac12}]_+\right]$

b) $\{H_n,H_m\} = 0$, all n,m.

Proof: In fact, the 1st part of (a) is just the lemma, and for the 2nd,

$$\begin{aligned}[D^2+q,(D^2+q)_+^{n-\frac12}] &= -[D^2+q,(D^2+q)_{\le -1}^{n-\frac12}]\\ &= -[D^2+q,\ (D^2+q)_{-1}^{n-\frac12}]_+\\ &= -[D^2+q,\ D^{-1}\cdot f_n(q,\dot q,\cdots)]_+\\ &= -(D\circ f_n - D^{-1}f_n D^2)_+\\ &= -(\dot f_n+\dot f_n)\\ &= -2g_n.\end{aligned}$$

As for (b)

$$\{H_n, H_m\}_{D^2+q} = \langle V_{H_n}, dH_m\rangle_{D^2+q}$$

$$= \langle -(n+\tfrac{1}{2})\,[D^2+q, (D^2+q)_+^{n-\frac{1}{2}}],\ (m+\tfrac{1}{2})\,(D^2+q)_{deg-1}^{n-\frac{1}{2}}\rangle$$

$$= -(n+\tfrac{1}{2})\,(m+\tfrac{1}{2})\,\mathrm{tr}\left([D^2+q, (D^2+q)_+^{n-\frac{1}{2}}]\circ(D^2+q)^{m-\frac{1}{2}}\right)$$

but

$$\mathrm{tr}([A,B]\circ C) = \mathrm{tr}(ABC-BAC) = \mathrm{tr}(B(CA-AC)) = 0 \text{ if } [A,C] = 0.$$

These flows V_{H_n} are the KdV dynamical system. Note that they are defined by universal polynomials $2g_n$ so in fact they make sense for any differential ring R:

$$(V_{H_1})_{D^2+q} = \dot{q}$$

(which integrates to q(x+t) i.e., it is just transl.)

$$(V_{H_2})_{D^2+q} = \left(\frac{6q\dot{q}+\dddot{q}}{4}\right)$$

$$(V_{H_3})_{D^2+q} = \left(\frac{30q^2\dot{q}+10q\dddot{q}+20\dot{q}\ddot{q}+q^{(5)}}{16}\right)$$

etc.

We want to elaborate on the conclusions that we have drawn. First of all, notice that combining the last Theorem with Corollary 11.15, we have reproven the conclusion of §10:

$$\begin{cases} \text{for all } k,\ell, \text{ there is a polynomial } H_{k,\ell}(q,\dot{q},\cdots) \text{ such that} \\ \dot{H}_{k\ell} = F_k G_\ell \end{cases}$$

Alternatively, we could have used this to prove $\{H_i, H_j\} = 0$. Secondly, notice that the conclusion

$$[V_{H_i}, V_{H_j}] = 0$$

makes sense over any differential ring R even when the H_i don't. Namely the vector fields V_{H_i} may be defined by part (a) of Theorem 11.18, and their commutativity may be expressed by (11.12)' by the polynomial identity

$$\sum_{k=0}^{\infty} \frac{\partial g_i}{\partial q^{(k)}} \cdot (g_j)^{(k)} = \sum_{k=0}^{\infty} \frac{\partial g_j}{\partial q^{(k)}} \cdot (g_i)^{(k)}.$$

In particular, over the ring R_1 of analytic functions, it follows that starting with any analytic function f(x) defined by $|x| < \varepsilon$, we can integrate any finite set of the flows V_{H_i}, getting an analytic $f(x,t_1,\cdots,t_n)$ defined for $|x|,|t_1|,\cdots,|t_n| < \eta$ such that

$$\frac{\partial f}{\partial t_1} = \dot{f}$$

$$\frac{\partial f}{\partial t_2} = \frac{6f\cdot\dot{f}-f^{(3)}}{4}$$

$$\cdots\cdots\cdots$$

$$\frac{\partial f}{\partial t_i} = 2g_i(f,\dot{f},\cdots), \qquad 1 \le i \le n.$$

The seond form of these equations given in Theorem 11.18 is called a Lax equation. In general, if S is a vector space of operators X, and $t \longmapsto X(t)$ is a 1-parameter of operators and $\Phi: S \longrightarrow S$ is a way of transforming one operator into another, then a Lax equation for the family X(t) is an equation:

$$\frac{d}{dt}X(t) = [X(t),\Phi(X(t))].$$

The importance of such equations is that they say that the operators X(t) are infinitesimally conjugate to each other, i.e.,

$$X(t_0+\delta t) = (I+\delta t\cdot\Phi(X(t_0)))^{-1}\circ X(t_0)\circ(I+\delta t\cdot\Phi(X(t_0))) \mod \delta t^2$$

In good cases, this implies that any sort of spectrum of X(t) is independent of t.

It is evident that this whole collection of flows on M mirrors the flows on Jac C, as defined in §10. The precise link is this:

Theorem 11.19: For any genus g, let C be a smooth hyperelliptic curve of genus g, and let $B,U,\infty,\wp$ be defined as usual. Let the vector field D on Jac C be written $\sum e_i\, \partial/\partial z_i$. Define an embedding

$$\begin{array}{ccc} \text{Jac } C - \Theta & \longrightarrow & M \\ \vec{z}_0 & \longmapsto & \left(\text{the operator } (\tfrac{d}{dx})^2+2\wp(\vec{z}_0+x\vec{e})\right) \end{array}$$

Then all the flows V_{H_n} on M are tangent to the image and V_{H_n} restricts to the flow $4^{-(n-1)}E_n$ on Jac C.

Proof: We simply combine the results of §10 and §11. Note that

$$E_k(2\wp) = \frac{G_k(\wp,\dot{\wp},\cdots)}{2} = 2\cdot{}^{k-1}g_k(2\wp,2\wp,\cdots) = 4^{k-1}(V_{H_k})_{D^2+2\wp}.$$

Corollary 11.20. At the point $(\frac{d}{dx})^2+2\wp(\vec{z}_0+x\vec{e}) \in M$, the vectors V_{H_n} span a finite-dimensional space of dimension g. In terms of the moduli α_i of C defined in §10, for all $k > g$

$$(V_{H_k})_{D^2+2\wp} + \frac{\alpha_1}{4}(V_{H_{k-1}})_{D^2+2\wp} + \cdots\cdots + \frac{\alpha_{k-1}}{4^{k-1}}(V_{H_1})_{D^2+\wp} = 0 .$$

Proof: Combine Theorem 11.19 and (10.7).

Corollary 11.21. For all $C, \vec{z}_0$, there is a differential operator of degree 2g+1 which commutes with $D^2+2\wp$, namely

$$\sum_{\ell=1}^{g+1} \alpha_{g+1-\ell}\cdot 4^{\ell-g-1}(D^2+2\wp)_+^{\ell-\frac{1}{2}} .$$

Proof: Combine Cor. 11.20 and Theorem 11.18a.

One case where the KdV dynamical system has been explored much more deeply, first by McKean-Van Moerbeke, then by McKean-Trubowitz, is over the ring

$$R_4 = C^\infty \text{ periodic real functions on } \mathbb{R}.$$

We sketch their theory very briefly. The operators

$$X_q = (\frac{d}{dx})^2 + q(x)$$

with q periodic can be analyzed by the Floquet theory (cf. Magnus, Hill's equation). In particular, for all $h \in \mathbb{C}^*$, they have a so-called h-spectrum:

$$\text{h-spectrum} = \left\{ \begin{array}{l} \text{set of } \lambda \text{ s.t. there is an eigenfunction} \\ f(x) \text{ with } \quad f''(x)+q(x)\cdot f(x) = \lambda f(x) \\ \qquad\qquad\qquad f(x+1) = hf(x) \end{array} \right\}$$

The fact that the K-dV flows can be written in the Lax form

$$\frac{\partial q_t}{\partial t} = \left[\left(\frac{d}{dx}\right)^2 + q_t(x), \left(\left(\frac{d}{dx}\right)^2 + q_t(x)\right)_+^{n-\frac{1}{2}} \right]$$

or

$$\frac{\partial}{\partial t} X_{q_t} = [X_{q_t}, Y_{q_t}]$$

shows by standard results that the h-spectrum is constant as a function of t. We may now consider

$$\Sigma(q) = \{(h,\lambda) \mid \lambda \in \text{h-spectrum}\} \subset \mathbb{C}^2$$

which is readily seen to be a 1-dimensional complex analytic subset such that the projection $\Sigma(q) \longrightarrow (\lambda\text{-plane})$ is 2-1. In fact, for each λ,

let $\left.\begin{array}{l} f_0(x,\lambda) \\ f_1(x,\lambda) \end{array}\right\}$ be the 2 solutions of $f''+qf = \lambda f$ with

$$\begin{cases} f_0(0,\lambda) = 1, & f_0'(0,\lambda) = 0 \\ f_1(0,\lambda) = 0, & f_1'(0,\lambda) = 1. \end{cases}$$

Then $\left.\begin{array}{l} f_0(x+1,\lambda) \\ f_1(x+1,\lambda) \end{array}\right\}$ are again 2 solutions, so we can write them

$$f_0(x+1,\lambda) = a(\lambda)f_0(x,\lambda)+b(\lambda)f_1(x,\lambda)$$

$$f_1(x+1,\lambda) = c(\lambda)f_0(x,\lambda)+d(\lambda)f_1(x,\lambda)$$

a,b,c,d entire analytic functions of λ such that $ad-bc \equiv 1$.

Then $\Sigma(q)$ is defined by

$$h^2-(a(\lambda)+d(\lambda))h + 1 = 0.$$

Let $\Sigma^*(q)$ be the normalization of $\Sigma(q)$, i.e., the 2 sheets separated at double zeroes of the discriminant $\Delta(\lambda) = (a+d)^2-4$ if any. Thus a "hyperelliptic curve" $\Sigma^*(q)$ usually of infinite genus is associated to this situation. The basic result of the theory of McKean and collaborators is that for all $\Sigma^*(q) \longrightarrow \mathbb{C}$ of finite or infinite genus the following sets are equal:

1) $\{q_1 \mid$ the branch points of $\Sigma^*(q_1) \longrightarrow \mathbb{C}$, $\Sigma^*(q) \longrightarrow \mathbb{C}$ are the same, hence $\Sigma^*(q_1) \cong \Sigma^*(q)\}$
2) $\{q_1 \mid \Sigma(q_1) = \Sigma(q)$ as subsets of $\mathbb{C}^2\}$
3) the orbit of the KdV flows through q
4) the set of all q_1 such that the KdV Hamiltonians $\int_0^1 F_n(q_1,\dot{q}_1,\cdots)dx = \int_0^1 F_n(q,\dot{q},\cdots)$ are equal.

Moreover, this set is canonically isomorphic to a distinguished component of the subgroup of real points on the Jacobian of C.

§1. The Prime Form $E(x,y)$.

Given an arbitrary compact Riemann surface X, of genus g, wouldn't it be handy if we had a holomorphic function $E: X \times X \longrightarrow \mathbb{C}$ such that $E(x,y) = 0$ if and only if $x = y$? Although such a function doesn't exist, it turns out that it "almost" does! To understand part of the problem and how to fix it, let's look at the simplest case:

Example. Let $X = \mathbb{P}^1$. The function $x-y$ works on $\mathbb{P}^1-\{\infty\}$ but not on all of $\mathbb{P}^1$. So consider instead the "differential":

$$E(x,y) = \frac{x-y}{\sqrt{dx}\ \sqrt{dy}},$$

where $\sqrt{dx}, \sqrt{dy}$ are defined as follows:

Choose a line bundle square root L of $\Omega^1_{\mathbb{P}^1}$ and an isomorphism $L^{\otimes 2} \xrightarrow{\sim} \Omega^1_{\mathbb{P}^1}$. Choose a section $\sqrt{dx} \in \Gamma(\mathbb{P}^1-\{\infty\},L)$ such that $(\sqrt{dx})^2 = dx$ (under this isomorphism). To check that this is finite on all of $\mathbb{P}^1 \times \mathbb{P}^1$, let $x' = 1/x$ be a coordinate on $\mathbb{P}^1-\{0\}$. Then

$$dx' = -\frac{1}{x}dx$$

so define $\sqrt{dx'}$ by

$$\sqrt{dx'} = \frac{\sqrt{-1}}{x}\sqrt{dx}\ .$$

Then if $x' = 1/x$, $y' = 1/y$,

$$E(x,y) = \frac{x-y}{\sqrt{dx}\,\sqrt{dy}} = \frac{\frac{1}{x'}-\frac{1}{y'}}{-\frac{\sqrt{dx'}}{x'}\cdot\frac{\sqrt{dy'}}{y'}} = \frac{x'-y'}{\sqrt{dx'}\,\sqrt{dy'}} .$$

For a general compact Riemann surface X, we will have to modify this approach in several ways. First, choose an L and an isomorphism $L^{\otimes 2} \xrightarrow{\sim} \Omega_X^1$ such that $h^0(L) = 1$. In terms of divisors, this means we want to find a divisor class $\delta \in \mathrm{Pic}^{g-1}(X)$ such that a) $2\delta \sim K_X$ and

b) $|\delta|$ = a single divisor δ .

We must show that such a δ exists:

Lemma 1. $\exists\ \delta \in \mathrm{Pic}^{g-1}(X)$ as above, i.e., there exists a nonsingular, odd, theta characteristic.

Proof. We use Riemann's theorem, and Lefschetz' embedding theorem. We want to translate the conditions on δ into theta functions:

$$\delta \text{ exists} \iff \exists\, [{\delta' \atop \delta''}] \in \tfrac{1}{2}\mathbb{Z}^{2g}/\mathbb{Z}^{2g} \quad \text{s.t.}$$

$$\text{a)}\quad \vartheta[{\delta' \atop \delta''}](o,\Omega) = 0$$

$$\text{and}\quad \text{b)}\quad d_z\vartheta[{\delta' \atop \delta''}](o,\Omega) \neq 0 .$$

Now, Lefschetz' theorem states that:

$$\mathbb{C}^g/2L_\Omega \longrightarrow \mathbb{P}^{(2^g-1)}$$

$$z \longmapsto (\ldots,\vartheta[{\delta' \atop \delta''}](z,\Omega),\ldots) \quad , \ \delta',\delta'' \in \tfrac{1}{2}\mathbb{Z}^g/\mathbb{Z}^g$$

is an embedding. In particular, the differentials $d_z\vartheta[{\delta' \atop \delta''}](o,\Omega)$ must span the cotangent space. Note that $\vartheta[{\delta' \atop \delta''}](o,\Omega) \neq 0$ implies $[{\delta' \atop \delta''}]$ is even, hence $\vartheta[{\delta' \atop \delta''}](z,\Omega)$ is invariant under $z \longmapsto -z$, hence $d_z\vartheta[{\delta' \atop \delta''}](o,\Omega) = 0$. Thus if $[{\delta' \atop \delta''}]$ satisfies (b), it also satisfies (a). QED

By Riemann's theorem, let $\zeta = \sum_{i=1}^{g} \frac{\partial\vartheta[{\delta' \atop \delta''}]}{\partial z_i}(o)\omega_i$ be the unique 1-form which is zero on δ, where $[{\delta' \atop \delta''}]$ corresponds as above to δ. In fact, $(\zeta) = 2\delta$, since $(\zeta)-\delta$ = (an effective divisor in $K-\delta$, i.e., δ). So $\zeta = (\sqrt{\zeta})^2$, where $\sqrt{\zeta}$ is a section of L. We may think of $\sqrt{\zeta}$ as a differential form of weight $\frac{1}{2}$. This $\sqrt{\zeta}$ will take the place of $\sqrt{dx}$.

Next we modify the numerator x-y in E, using a theta function $\vartheta[{\delta' \atop \delta''}])(\int_x^y \vec{\omega})$ for higher genus.

Definition. The prime form $E(x,y)$ is given by

$$E(x,y) = \frac{\vartheta[{\delta' \atop \delta''}](\int_x^y \vec{\omega})}{\sqrt{\zeta(x)}\ \sqrt{\zeta(y)}}$$

where: a) δ is a fixed nonsingular odd theta characteristic.

b) δ corresponds to $[{\delta' \atop \delta''}]$.

c) $\sqrt{\zeta(x)}$, $\sqrt{\zeta(y)}$ are as above.

This is a holomorphic differential form of weight $(-\frac{1}{2}, -\frac{1}{2})$ on $\tilde{X} \times \tilde{X}$, where $\tilde{X}$ is the universal cover of X.

A few remarks:

1. E is not defined on $X \times X$ since a choice of path of integration from x to y must be made. To make this well-defined we simply pull back to the universal cover.

2. Note, however, that whether this is zero or not only depends on the image of x,y in X: $E(\tilde{x}_1,\tilde{y}_1) = 0 \iff E(\tilde{x}_2,\tilde{y}_2) = 0$ when $\tilde{x}_1,\tilde{x}_2$, resp. $\tilde{y}_1,\tilde{y}_2$ have the same projection to X. Alternatively, we can consider E(x,y) as a holomorphic section of a line bundle on $X \times X$.

The following properties make the prime form useful:

Properties of E(x,y). Let $x,y \in \tilde{X}$, $\bar{x},\bar{y}$ their images in X.

1. $E(x,y) = 0 \iff \bar{x} = \bar{y}$. [This is its major property.]
2. E has a first order zero along the diagonal $\Delta \subset X \times X$.
3. $E(x,y) = -E(y,x)$.
4. Choose a local coordinate t about $x \in X$ such that $\zeta = dt$. Then
$$E(x,y) = \frac{t(x) - t(y)}{\sqrt{dt(x)}\,\sqrt{dt(y)}}\,(1 + O(t(x)-t(y))^2).$$
5. If x or y is moved by an A-period, E(x,y) remains invariant.
If x is moved by a B-period $\Sigma m_i B_i$ to x',
$$E(x,y) = \pm E(x,y)\exp(-\pi i\, {}^t m\Omega m + 2\pi i\, {}^t m \int_x^y \vec{\omega}).$$
If y is similarly moved to y':
$$E(x,y') = \pm E(x,y)\exp(-\pi i\, {}^t m\Omega m - 2\pi i\, {}^t m \int_x^y \vec{\omega}).$$

The main lemma that we need to prove this is:

<u>Lemma 2</u>. <u>Given</u> δ <u>as above</u>, $|\delta| = \left\{\sum_{i=1}^{g-1} P_i\right\}$, <u>then</u>

$$\vartheta[{}^{\delta'}_{\delta''}]\left(\int_x^y \vec{\omega}\right) = 0 \iff \begin{array}{ll} \text{a)} & x = y \quad \underline{\text{or}} \\ \text{b)} & x = \underline{\text{some}}\ P_i\ \underline{\text{or}} \\ \text{c)} & y = \underline{\text{some}}\ P_i. \end{array}$$

<u>Proof</u>. By Riemann's theorem,

$$\vartheta[{}^{\delta'}_{\delta''}]\left(\int_x^y \vec{\omega}\right) = 0 \iff |y-x+\delta| \neq \emptyset.$$

Now $h^0(\Sigma P_i) = 1$, so $h^0(y + \Sigma P_i) = 1$ or 2.

<u>Case 1</u>. $h^0(y + \Sigma P_i) = 1$.

So $|y + \Sigma P_i| = \{y + \Sigma P_i\}$

$$|y - x + \Sigma P_i| \neq \emptyset \iff \text{either } x = y, \text{ or } x = \text{some } P_i.$$

<u>Case 2</u>. $h^0(y + \Sigma P_i) = 2$.

By Riemann-Roch, $h^0(K - y - \Sigma P_i) = 1$, but

$$K - \Sigma P_i \sim \Sigma P_i$$

so $h^0(\Sigma P_i - y) = 1$

so $y = \text{some } P_i$. <u>QED</u>

The proofs of the properties above are now quite easy. For instance, for (1): From the lemma we know:

$$V\left(\vartheta[{}^{\delta'}_{\delta''}]\left(\int_x^y \vec{\omega}\right)\right) = V(x-y) \cup \left(\bigcup_i P_i \times X\right) \cup \left(\bigcup_i X \times P_i\right).$$

The fact that it vanishes to order one is left to the reader. But this is precisely why we divided by $\sqrt{\zeta(x)}$, $\sqrt{\zeta(y)}$:

$$(\sqrt{\zeta(x)})_{\text{divisor}} = \sum P_i$$

so $(E(x,y)) = \Delta$ as divisors. For the others, (3), (5) are immediate, and (4) is just a local calculation. QED

As one application of the prime form, we will construct all meromorphic functions on X, as well as the basic differentials:

(a) Given $a_1,\ldots,a_n,b_1,\ldots,b_n \in X$ such that $\Sigma a_j \sim \Sigma b_i$ in Pic X and suppose $a_i \neq b_j$ for all i,j. Then

$$f(x) = \prod_{i\neq 1}^{n} \frac{E(x,a_i)}{E(x,b_i)}$$ is a single-valued meromorphic function

on X with zeros = Σa_i, poles = Σb_i.

To prove this, note that all the $\sqrt{\zeta(x)}$, etc., cancel out, so you are left with $\prod_{i=1}^{n} \frac{\vartheta[{\delta' \atop \delta''}](\int_x^{a_i}\vec{\omega})}{\vartheta[{\delta' \atop \delta''}](\int_x^{b_i}\vec{\omega})}$ and now just check invariance under A,B periods.

(b) Construction of differentials of the 3rd kind.

We want $\omega_{a-b}(x)$ = the unique differential 1-form on X with

a) zero A-periods

b) single pole at a with residue 1

single pole at b with residue -1.

In fact, $\omega_{a-b}(x) = d_x \log \frac{E(x,a)}{E(x,b)}$. To check this, look locally:

$$\omega_{a-b}(x) = d_x \log(t(x)-t(a)) - d_x \log(t(x)-t(b)) + \text{holomorphic differential}$$
$$= \frac{dt(x)}{t(x)-t(a)} - \frac{dt(x)}{t(x)-t(b)} + \text{holomorphic differential.}$$

(c) Construction of differentials of the 2nd kind.

We want $\eta_a(x)$ = a 1-form on X with

a) zero A-periods

b) double pole at $a \in X$.

Note that such an η_a is unique up to a multiplicative constant.

Consider:

$$\omega(x,y) = d_x d_y \log E(x,y).$$

This is a well-defined 2-form on $X \times X$, since $d_x\, d_y \log[f(x,y)\cdot g(x)\cdot h(y)] = d_x\, d_y \log f(x,y)$. For each fixed $y = a$, by choosing a basis for the tangent space to X at a, it restricts to a 1-form on X equal to the above $\eta_a(x)$ up to a multiplicative constant. In this manner, we can construct differential 1-forms with any allowed divisor of zeros and poles.

§2. Fay's Trisecant Identity

We now come to a very fundamental identity between theta functions that holds for the period matrices of curves, but not for general period matrices. Although the basic ideas behind this identity go back to Riemann, it was not clearly isolated until Fay made his beautiful and systematic analysis of the theory of theta functions (J. Fay, Theta functions on Riemann surfaces, Springer Lecture Notes 352, 1973).

Theorem [Fay, *op. cit.*, p. 34, formula 45]. *Let* X *be a compact Riemann surface,* $\tilde{X}$ *its universal covering space,* $\vartheta(\vec{z})$ *its associated theta function and* $E(x,y)$ *its prime form. Then for all* $a,b,c,d \in \tilde{X}$, $\vec{z} \in \mathbb{C}^g$:

$$\vartheta\left(\vec{z} + \int_a^c \vec{\omega}\right).\vartheta\left(\vec{z} + \int_b^d \vec{\omega}\right).E(c,b)E(a,d)$$

$$+\vartheta\left(\vec{z} + \int_b^c \vec{\omega}\right).\vartheta\left(\vec{z} + \int_a^d \vec{\omega}\right).E(c,a)E(d,b)$$

$$= \vartheta\left(\vec{z} + \int_{a+b}^{c+d} \vec{\omega}\right).\vartheta(\vec{z}).E(c,d)E(a,b).$$

This type of identity is very special. The theta function on general abelian varieties doesn't satisfy identities like $\sum_{i=1}^{3} c_i\,\vartheta(z+a_i).\vartheta(z+b_i) = 0.$

The proof of the theorem falls into several steps, each of which is straightforward but sometimes tedious.

Step I. Check that all three terms satisfy the same functional equations and are differentials of the same type. This way all three terms will be sections of the same line line bundle L on the space $X \times X \times X \times X \times \mathrm{Jac}(X)$.

Step II. Next show that if both terms on the left are zero, then the right hand side is also zero.

Step III. Let D_1, D_2 be the codimension one subsets where the two terms on the left, respectively, are zero. Then for all components D_3 of $D_1 \cap D_2$, the intersection is generically transversal.

Step IV. $H^1(X \times X \times X \times X \times \mathrm{Jac}(X), L^{-1}) = 0$.

Step V. Assume: X smooth complete variety

L line bundle on X such that $H^1(X, L^{-1}) = 0$.

$t, s_1, s_2 \in H^0(X, L)$ global sections s.t.

a) $s_1 = 0$, $s_2 = 0$ are divisors D_1, D_2 without multiplicity.

b) For all components D, of D_1 D_2, the intersection is generically transversal.

c) $s_1(x) = s_2(x) = 0 \Longrightarrow t(x) = 0$.

Then: $\exists \lambda_1, \lambda_2$ such that

$$t = \lambda_1 s_1 + \lambda_2 s_2.$$

Step VI. First, let a = b, secondly let b = c, to see that the constants are one, finishing the proof.

We will not go through all the details but instead touch on all the main points:

Step I: Everything is invariant under $\vec{z} \longmapsto \vec{z}+\vec{n}$, $\vec{n} \in \mathbb{Z}^g$. Under $\vec{z} \longmapsto \vec{z} + \Omega\vec{m}$, $\vec{m} \in \mathbb{Z}^g$, the 3 terms are multiplied by

$$e^{-\pi i\, {}^t m\Omega m - 2\pi i\, {}^t m\,(z+\int_a^c \omega)} \cdot e^{-\pi i\, {}^t m\Omega m - 2\pi i\, {}^t m(z+\int_b^d \omega)},$$

$$e^{-\pi i\, {}^t m\Omega m - 2\pi i\, {}^t m(z+\int_b^c \omega)} \cdot e^{-\pi i\, {}^t m\Omega m - 2\pi i\, {}^t m(z+\int_a^d \omega)},$$

and

$$e^{-\pi i\, {}^t m\Omega m - 2\pi i\, {}^t m(z+\int_{a+b}^{c+d} \omega)} \cdot e^{-\pi i\, {}^t m\Omega m - 2\pi i\, {}^t m\cdot z},$$

respectively, which are equal because

$$\int_a^c \omega + \int_b^d \omega = \int_b^c \omega + \int_a^d \omega = \int_{a+b}^{c+d} \omega$$

on $(\tilde{X})^4$. Among the many substitutions in a,b,c,d, we consider only $c \longmapsto \gamma c$, $\gamma \in \pi_1(x)$. Let $\vec{n}+\Omega\vec{m}$ be the period defined by γ. Note that the half-order differentials $\sqrt{\zeta(x)}$ are sections of a line bundle on X, hence are invariant by all such substitutions. Thus

$$E(\gamma c,b) = e^{-\pi i\, {}^t m\Omega m} \cdot e^{-2\pi i\, {}^t m(\int_b^c \omega + \delta'') - 2\pi i\, {}^t n\cdot\delta'} \cdot E(c,b).$$

Collecting all the factors, you find that <u>all</u> 3 terms are multiplied by

$$e^{-2\pi i\, {}^t m\Omega m - 2\pi i\, {}^t m\delta'' - 2\pi i\, {}^t n\cdot\delta' - 2\pi i\, {}^t m(z+\int_{a+b}^{2c} \omega)}.$$

Step II: One must look at all 16 combinations of one of the 4 factors of the 1st term with one of the 4 factors of the 2nd term. Most combinations are obvious, e.g.,

$$E(a,d) = 0,\ E(c,a) = 0 \Longrightarrow E(c,d) = 0$$

$$\theta(z + \int_b^d \omega) = 0,\ E(c,a) = 0 \Longrightarrow \theta(z + \int_{a+b}^{c+d} \omega) = 0 .$$

A slightly less obvious case is when

$$\theta(z + \int_a^c \omega) = 0, \quad \theta(z + \int_b^c \omega) = 0.$$

If D_z is the divisor of degree g-1 on X defined by z, this means

$$|D_z + c - a| \neq \emptyset$$
$$|D_z + c - b| \neq \emptyset .$$

Then either $a = b$, or $|D_z+c|$ is a pencil or $|D_z+c-a-b| \neq \emptyset$.
Therefore either $a = b$, or $|D_z| \neq \emptyset$ or $|D_z+c+d-a-b| \neq \emptyset$.
Therefore either $E(a,b) = 0$ or $\theta(z) = 0$ or $\theta(z + \int_{a+b}^{c+d} \omega) = 0$.

Step III: Let's look at the generic transversality of $\theta(z + \int_a^c \omega) = 0$ and $\theta(z + \int_b^c \omega) = 0$. We can ignore loci of codimension >2. Recall that the differential $d_z\theta$ vanishes at $z = z_o$ if and only if $|D_{z_o}|$ is a pencil; and if $|D_{z_o}|$ is a single divisor F_{z_o}, this differential pulls back on X to the unique 1-form ω_{z_o} zero on F_{z_o}. Thus the loci where

$$d_z\theta(z + \int_a^c \omega) = 0 \quad \text{or} \quad d_z\theta(z + \int_b^c \omega) = 0$$

are the loci where $|D_z+c-a|$ or $|D_z+c-b|$ are pencils: these have higher codimension and can be ignored. We may also suppose that of the 3 alternatives: (1) $a = b$, (2) $|D_z+c|$ pencil and (3) $|D_z+c-a-b| \neq \emptyset$, exactly one holds. Let ω_a, resp. ω_b, be the 1-forms on X zero on the divisor in $|D_z+c-a|$, resp. $|D_z+c-b|$. If $\omega_a \neq \omega_b$, this means that the differentials of $\theta(z + \int_a^c \omega)$ and $\theta(z + \int_b^c \omega)$ in the z-direction are independent. If $a = b$, but $\omega_a(a) \neq 0$, this means that in the a-direction $\theta(z + \int_a^c \omega)$ has non-zero differential while $\theta(z + \int_b^c \omega)$ has zero differential. In both cases, the intersection is transversal.

But now there are 3 possibilities

<u>Case 1</u>: $a = b$, $|D_z+c|$ is one divisor, $|D_z+c-a-b| = \emptyset$.

<u>Case 2</u>: $a \neq b$, $|D_z+c|$ pencil, $|D_z+c-a-b| = \emptyset$.

<u>Case 3</u>: $a \neq b$, $|D_z+c|$ one divisor, $|D_z+c-a-b| \neq \emptyset$.

It is not hard to show that in case 1, $\omega_a(a) \neq 0$ while in cases 2 and 3, $\omega_a \neq \omega_b$.

<u>Step IV</u>: Look at the projection

$$\begin{array}{c} X \times X \times X \times X \times J \\ \Big\downarrow p_{345} \\ X \times X \times J \end{array}$$

where a,b vary in the fibres. L restricts to each fibre $p_{345}^{-1}(z)$

to a line bundle of the form $p_1^{-1}(M_1) \otimes p_2^{-1}(M_2)$ where M_1, M_2 have positive degree. By the Künneth formula, $H^i(L^{-1}|_{\text{fibre}}) = (0)$, $i = 0,1$, hence by the Leray spectral sequence $H^1(L^{-1}) = (0)$.

Step V: Use the exact sequence

$$0 \longrightarrow L^{-1} \xrightarrow[(s_1,s_2)]{} \mathcal{O}_X \oplus \mathcal{O}_X \xrightarrow[\binom{s_2}{-s_1}]{} I_{\mathfrak{Z}} \cdot L \longrightarrow 0$$

where $\mathfrak{Z}$ is the subscheme $s_1 = s_2 = 0$.

Step VI: Obvious.

Next, we want to give a geometric interpretation of the identity.

In fact:

a) use $|2\Theta|$ to map $\mathrm{Jac}(X)$ to projective space. The image is called the Kummer variety.

b) in this mapping, the trisecant identity will tell us that the images of certain sets of three points (∞^4 of them!) are collinear, i.e., the "Kummer variety" has ∞^4 trisecants.

First, we need to know what $|2\Theta|$ consists of:

Lemma.

$$|2\Theta| = \left\{ \begin{array}{l} \text{the set of divisors of the form} \\ \left(\sum_{\eta \in \frac{1}{2}\mathbb{Z}^g/\mathbb{Z}^g} c_\eta \vartheta\left[\begin{smallmatrix}0\\ \eta\end{smallmatrix}\right](z,\tfrac{\Omega}{2}) = 0 \right), \quad \text{for all } (c_\eta) \in \mathbb{C}^{2^g} \end{array} \right\}$$

Proof. First, it is easy to check that $\left(\vartheta\left[\begin{smallmatrix}0\\ \eta\end{smallmatrix}\right](z,\tfrac{\Omega}{2}) = 0\right) \in |2\Theta|$:

We have $\vartheta[{}^{0}_{\eta}](z+n, \frac{\Omega}{2}) = \vartheta[{}^{0}_{\eta}](z, \frac{\Omega}{2})$

and $\vartheta[{}^{0}_{\eta}](z+\Omega m, \frac{\Omega}{2}) = e^{-4\pi i\, {}^{t}\eta n}\, e^{-2\pi i\, {}^{t}m\Omega m - 4\pi i\, {}^{t}mz}\, \vartheta[{}^{0}_{\eta}](z, \frac{\Omega}{2})$.

Since the transition functions are the squares of those of $|\Theta|$,

$$V(\vartheta[{}^{0}_{\eta}](z, \tfrac{\Omega}{2})) \in |2\Theta| .$$

Next, we must check that these span $|2\Theta|$. One way is to find the dimension of $H^{0}(2\Theta)$: it will be 2^{g}. Since the $\vartheta[{}^{0}_{\eta}](z,\frac{\Omega}{2})$ are all linearly independent, this shows they are a basis. To find the dimension, let $f(z) = \sum_{n\in\mathbb{Z}^{g}} a(n)e^{2\pi i\, {}^{t}nz} \in H^{0}(2\Theta)$. Thus:

$$f(z+\Omega m) = e^{-2\pi i\, {}^{t}m\Omega m - 4\pi i\, {}^{t}mz} f(z) .$$

This gives us a formula for $a(k)$:

$$a(k+2m) = a(k) \cdot e^{-2\pi i\, {}^{t}m\Omega m} \cdot k,m \in \mathbb{Z}^{g} .$$

This gives us an upper bound of 2^{g} for the dimension, which is what we wanted. QED

Moreover, recall from Chapter II, the fundamental:

Addition Theorem.

$$\vartheta(x+y,\Omega)\vartheta(x-y,\Omega) = 2^{-g} \sum_{\eta\in\frac{1}{2}\mathbb{Z}^{g}/\mathbb{Z}^{g}} \vartheta[{}^{0}_{\eta}](x, \tfrac{\Omega}{2})\vartheta[{}^{0}_{\eta}](y,\tfrac{\Omega}{2}) .$$

The geometric interpretation of the theorem is

Geometric Corollary. Let $|2\Theta|$ define $\phi\colon \mathrm{Jac}(X) \longrightarrow \mathbb{P}^{(2^g-1)}$. Then, $\forall a,b,c,d \in \tilde{X}$, the three points

$$\phi\left(\frac{1}{2}\int_{a+b}^{c+d} \vec{\omega}\right), \quad \phi\left(\frac{1}{2}\int_{b+d}^{a+c} \vec{\omega}\right), \quad \phi\left(\frac{1}{2}\int_{a+d}^{b+c} \vec{\omega}\right)$$

are collinear.

Proof. The map ϕ, by the lemma above, is given explicitly as

$$\phi(z) = \left(\ldots, \vartheta\left[\begin{smallmatrix}0\\ \eta\end{smallmatrix}\right](z,\tfrac{\Omega}{2}),\ldots\right)_{\eta\in\frac{1}{2}\mathbb{Z}^g/\mathbb{Z}^g} .$$

We want to use the trisecant identity to give us one relation.

Let V = vector space spanned by $\vartheta\left[\begin{smallmatrix}0\\ \eta\end{smallmatrix}\right](z, \frac{\Omega}{2})$, $\eta \in \frac{1}{2}\mathbb{Z}^g/\mathbb{Z}^g$.

Let Q = symmetric bilinear form on V with these $\vartheta\left[\begin{smallmatrix}0\\ \eta\end{smallmatrix}\right](z, \frac{\Omega}{2})$ as an orthonormal basis.

Now note,

1) For any $a \in \mathbb{C}^g$, $\vartheta(z+a,\Omega)\vartheta(z-a,\Omega) \in V$.

   (For instance, use the addition formula to get this.)

2) Let $T_a^*\vartheta(z,\Omega) = \vartheta(z+a,\Omega)$. Then,

$$\forall f \in V, \qquad Q(T_a^*\vartheta \cdot T_{-a}^*\vartheta,\ f) = 2^{-g}\cdot f(a).$$

   (Just check this on basis elements $f(z) = \vartheta\left[\begin{smallmatrix}0\\ \eta\end{smallmatrix}\right](z, \frac{\Omega}{2})$.
   By the addition theorem, both sides are $2^{-g}\vartheta\left[\begin{smallmatrix}0\\ \eta\end{smallmatrix}\right](a, \frac{\Omega}{2})$).

Now we can apply the trisecant theorem. In the theorem, make the substitution $z \longmapsto z - \frac{1}{2}\int_{a+b}^{c+d}\vec{\omega}$. We get:

$$\text{LHS} = c_1\vartheta(z + \tfrac{1}{2}\int_{a+d}^{b+c}\vec{\omega})\vartheta(z - \tfrac{1}{2}\int_{a+d}^{b+c}\vec{\omega})$$

$$+ c_2\vartheta(z + \tfrac{1}{2}\int_{b+d}^{a+c}\vec{\omega})\vartheta(z - \tfrac{1}{2}\int_{b+d}^{a+c}\vec{\omega})$$

$$\text{RHS} = c_3\vartheta(z + \tfrac{1}{2}\int_{a+b}^{c+d}\vec{\omega})\vartheta(z - \tfrac{1}{2}\int_{a+b}^{c+d}\vec{\omega}) .$$

Note that these three products of ϑ's are of the form of the function in notes 1,2. Let $f(z) = \vartheta[{}^{o}_{\eta}](z, \frac{\Omega}{2})$ for any $\eta \in \frac{1}{2}\mathbb{Z}^g/\mathbb{Z}^g$. Apply $Q(\_,f)$ to the equation to obtain

$$c_1\vartheta[{}^{o}_{\eta}](\tfrac{1}{2}\int_{a+d}^{b+c}\vec{\omega}, \tfrac{\Omega}{2}) + c_2\vartheta[{}^{o}_{\eta}](\tfrac{1}{2}\int_{b+d}^{a+c}\vec{\omega}, \tfrac{\Omega}{2})$$

$$= c_3\vartheta[{}^{o}_{\eta}](\tfrac{1}{2}\int_{a+b}^{c+d}\vec{\omega}, \tfrac{\Omega}{2})$$

where c_1, c_2, c_3 are independent of η. <u>QED</u>

§3. Corollaries of the identity

In this section we will study what happens to Fay's identities when the 4 points a,b,c,d come together in various stages. The result will be identities involving derivatives of theta functions. First, we need some notation. For the following formulas, let

a) $\vec{z} \in \mathbb{C}^g$

b) $a,b,c,d \in \tilde{X}$ with distinct projections to X

c) $\vartheta(\vec{z})$ the theta function of X

d) for every $a \in X$, and local coordinates t on X near a, we expand the differentials of the 1st kind:

$$\omega_i = \left(\sum_{j=0}^{\infty} v_{ij} \frac{t^j}{j!}\right) dt$$

and let

$$\vec{v}_j = (v_{1j}, \ldots, v_{gj}).$$

(Note that the mapping

$$\tilde{X} \longrightarrow \mathbb{C}^g$$
$$x \longmapsto \int_a^x \vec{\omega}$$

is given near a by

$$t \longmapsto \sum_{j=0}^{\infty} \vec{v}_j \frac{t^{j+1}}{(j+1)!}\ .)$$

We let

D_a = constant vector field $\vec{v}_0 \cdot \frac{\partial}{\partial \vec{z}}$ (i.e., $\sum v_{oi} \frac{\partial}{\partial z_i}$)

D'_a = constant vector field $\vec{v}_1 \frac{\partial}{\partial \vec{z}}$

D''_a = constant vector field $\vec{v}_2 \frac{\partial}{\partial \vec{z}}$.

e) We abbreviate $\int_a^b \vec{\omega}$ to $\int_a^b$.

The identities we will prove are:

(1) (Fay, Prop. 2.10, formula 38):

$$D_b \log \frac{\vartheta(\vec{z}+\int_a^c)}{\vartheta(\vec{z})} = c_1 + c_2 \frac{\vartheta(\vec{z}+\int_b^c)\,\vartheta(\vec{z}+\int_a^b)}{\vartheta(\vec{z}+\int_a^c)\cdot\vartheta(\vec{z})}$$

where $\omega_{a-c}(b) = c_1 dt$ (t a local coordinate near b)

$$\frac{E(a,c)}{E(b,c)E(a,b)} = c_2 dt .$$

(2) (Fay, Cor. 2.12, formula 38):

$$D_a D_b \log \vartheta(\vec{z}) = c_1 + c_2 \frac{\vartheta(\vec{z}+\int_a^b)\,\vartheta(\vec{z}+\int_b^a)}{\vartheta(\vec{z})^2}$$

where $\omega(a,b) = -c_1 dt_a\, dt_b$ (t_a, t_b local coord. near a,b resp.)

$$\frac{1}{E(a,b)^2} = c_2\, dt_a\, dt_b .$$

(3) (Fay, Cor. 2.13, p. 27):

$$D_a^4 \log \vartheta(z) + 6[D_a^2 \log \vartheta(z)]^2 - 2D_a D_a'' \log \vartheta(z)$$
$$+ 3 D_a'^2 \log \vartheta(z) + c_1 D_a^2 \log \vartheta(z) + c_2 = 0$$

where c_1, c_2 are constants depending on the Taylor expansions of $E(a,b)$ and $\omega(a,b)$.

As explained in Ch. II, there are 3 ways to get meromorphic functions on X_Ω from ϑ:

⟶ as products $\frac{\pi\vartheta(z+a_i)}{\pi\vartheta(z+b_i)}$ if $\sum a_i = \sum b_i$

⟶ as differences of logarithmic derivatives $D \log \frac{\vartheta(z+a)}{\vartheta(z)}$

⟶ as 2nd logarithmic derivatives $D D' \log \vartheta(z)$.

The above identities give basic identities between meromorphic functions formed in these 3 ways. Identities (1) will appear as the limiting case of the trisecant identity when $d \longrightarrow b$, while a,b,c are still distinct. Identity (2) will appear when in (1) we let $c \longrightarrow a$ while a,b are still distinct. Identity (3) will appear when finally we let $b \longrightarrow a$.

Before proving the formulas, we need the following lemma:

<u>Lemma</u>: a) $d_x \left. \frac{E(x,b)}{E(x,a)} \right|_{x=b} = \frac{1}{E(b,a)}$

b) $$\omega_{a-b}(x) = -\frac{E(x,a)}{E(x,b)}\, d_x \frac{E(x,b)}{E(x,a)}$$

c) $$d_x\, \omega_{a-x}(b)\Big|_{x=a} = -\omega(a,b)$$

(To prove (a), use the local expansion of E near x = b; (b) and (c) are restatements of the definitions.)

<u>Proof of formula one</u>:

We want:

(1b) $$D_b\vartheta\Big(z + \int_a^c\Big)\vartheta(z) - [D_b\vartheta(z)]\vartheta\Big(z + \int_a^c\Big)$$

$$= c_1\vartheta\Big(z + \int_a^c\Big)\vartheta(z) + c_2\vartheta\Big(z + \int_b^c\Big)\vartheta\Big(z + \int_a^b\Big).$$

Take the trisecant identity, and divide by E(c,b)E(a,d):

$$\vartheta\Big(z + \int_a^c\Big)\vartheta\Big(z + \int_b^d\Big) + \frac{E(c,a)E(d,b)}{E(c,b)E(a,d)}\vartheta\Big(z + \int_b^c\Big)\vartheta\Big(z + \int_a^d\Big)$$

$$= \frac{E(c,d)E(a,b)}{E(c,b)E(a,d)}\vartheta(z)\vartheta\Big(z + \int_{a+b}^{c+d}\Big).$$

Now differentiate w.r.t d (as scalar functions on our R.S.), and let $d \longrightarrow b$:

$$\vartheta\left(z + \int_a^c\right) \cdot D_b\vartheta(z) + \frac{E(c,a)}{E(c,b)}\, d_x \left.\frac{E(x,b)}{E(a,x)}\right|_{x=b} \vartheta\left(z + \int_b^c\right)\vartheta\left(z + \int_a^b\right)$$

$$= \frac{E(a,b)}{E(c,b)}\, d_x \frac{F(c,x)}{E(a,x)}\Bigg|_{x=b} \vartheta(z)\vartheta\left(z + \int_a^c\right)$$

$$+ \vartheta(z)\; D_b\vartheta\left(z + \int_a^c\right).$$

Now use the lemma (a),(b) to get:

$$\vartheta\left(z + \int_a^c\right) D_b\vartheta(z) + \frac{E(c,a)}{E(c,b)E(a,b)}\vartheta\left(z + \int_b^c\right)\vartheta\left(z + \int_a^b\right)$$

$$= -\omega_{a-c}(b)\vartheta(z)\vartheta\left(z + \int_a^c\right) + \vartheta(z) D_b\vartheta\left(z + \int_a^c\right).$$

This gives us our formula.

<u>Proof of formula two</u>:

We want:

(2b)
$$[D_a D_b\vartheta(z)]\,\vartheta(z) - D_a\vartheta(z)\cdot D_b\vartheta(z)$$
$$= c_1\,\vartheta(z)^2 + c_2\vartheta\left(z + \int_a^b\right)\vartheta\left(z + \int_b^a\right)$$

Now, take formula (1b), differentiate w.r.t. c, let c ⟶ a, while noticing that c_1, c_2 in (1b) are <u>not</u> constants w.r.t. c, and in fact both vanish when c = a:

$$[D_aD_b\vartheta(z)]\vartheta(z) - D_b\vartheta(z)\cdot D_a\vartheta(z)$$

$$= d_c\omega_{a-c}(b)\Big|_{c=a}\cdot\vartheta(z)^2$$

$$+ d_c\left(\frac{E(a,c)}{E(b,c)}\right)\Big|_{c=a}\cdot\frac{1}{E(a,b)}\vartheta\left(z+\int_b^a\right)\vartheta\left(z+\int_a^b\right)$$

But now use the lemma (a), (c), to get:

$$[D_aD_b\vartheta(z)]\vartheta(z) - D_b\vartheta(z)\cdot D_a\vartheta(z)$$

$$= -\omega(a,b)\,\vartheta(z)^2 + \frac{1}{E(a,b)^2}\,\vartheta\left(z+\int_b^a\right)\vartheta\left(z+\int_a^b\right),$$

which is exactly what we wanted.

<u>Proof of formula three:</u>

From (2) we have:

$$\text{(3b)}\quad E(a,b)^2[D_aD_b\log\vartheta(z)]\,\vartheta(z)^2 = -\omega(a,b)E(a,b)^2\vartheta(z)^2 + \vartheta\left(z+\int_a^b\right)\vartheta\left(z-\int_a^b\right).$$

The idea now is:

Let a,b be in the same coordinate neighborhood, $t(a) = 0$, $t(b) = t$

and expand (3b) in terms of t, and pick off the first non-trivial term, which will be the t^4 term!

(1) Locally we have:

$$\omega_i = \left(\sum_{j=0}^{\infty} v_{ij} \frac{t^j}{j!}\right) dt$$

$$\int_a^b \omega_i = v_{io} t + v_{i1} \frac{t^2}{2} + v_{i2} \frac{t^3}{6} + \cdots$$

$$D_b = \sum_{i=1}^{g} \left(\sum_{j=0}^{\infty} v_{ij} \frac{t^j}{j!}\right) \frac{\partial}{\partial z_i}$$

$$= D_a + t D_a' + \frac{t^2}{2} D_a'' + \cdots$$

$$E(a,b) = (t + c_1 t^3 + c_2 t^5 + \cdots) \cdot \frac{1}{\sqrt{dt(o)}\sqrt{dt}}$$

$$E(a,b)^2 = (t^2 + 2c_1 t^4 + (2c_2 + c_1^2)\, t^6 + \cdots) \frac{1}{dt(o)\ dt} \ .$$

Calculating from the definition, one easily checks:

$$\omega(a,b) = (\tfrac{1}{t^2} - 2c_1 + (6c_1^2 - 12c_2)t^2 + \cdots\cdots) dt(o)\ dt.$$

Hence

$$\omega(a,b) E(a,b)^2 = 1 + (3c_1^2 - 10c_2)t^4 + \cdots\cdots.$$

(2) Locally the term $E(a,b)^2\ D_a D_b \log \vartheta(z)$ is

$$= (t^2 + 2c_1 t^4 + \cdots)(D_a^2 \log \vartheta(z) + t\ D_a D_a' \log \vartheta(z) + \frac{t^2}{2} D_a D_a'' \log \vartheta(z) + \cdots)$$

$$= t^2 \cdot D_a^2 \log \vartheta(z) + t^3\ D_a D_a' \log \vartheta(z)$$

$$+ t^4 [2c_1\ D_a^2 \log \vartheta(z) + \tfrac{1}{2} D_a D_a'' \log \vartheta(z)] + \cdots .$$

(3) Let $\vartheta_i = \frac{\partial}{\partial z_i}\vartheta$, $\vartheta_{ij} = \frac{\partial^2}{\partial z_i \partial z_j}\vartheta$, etc. Then

$$\vartheta(z+\delta)\cdot\vartheta(z-\delta) = \Big[\vartheta(z) + \sum_{i=1}^{g}\delta_i\vartheta_i(z) + \frac{1}{2}\sum_{i,j=1}^{g}\delta_i\delta_j\vartheta_{ij}(z)$$

$$+ \frac{1}{6}\sum_{i,j,k=1}^{g}\delta_i\delta_j\delta_k\vartheta_{ijk}(z) + \cdots\Big]$$

$$\cdot\Big[\vartheta(z) - \sum_{i=1}^{g}\delta_i\vartheta_i(z) + \frac{1}{2}\sum_{i,j=1}^{g}\delta_i\delta_j\vartheta_{ij}(z)$$

$$- \frac{1}{6}\sum_{i,j,k=1}^{g}\delta_i\delta_j\delta_k\vartheta_{ijk}(z) + \cdots\Big]$$

$$= \vartheta(z)^2 + \sum_{i,j=1}^{g}\delta_i\delta_j\big(\vartheta(z)\vartheta_{ij}(z) - \vartheta_i(z)\vartheta_j(z)\big)$$

$$+ \sum_{i,j,k,\ell=1}^{g}\delta_i\delta_j\delta_k\delta_\ell\Big(\frac{1}{12}\vartheta(z)\vartheta_{ijk\ell}(z) - \frac{1}{3}\vartheta_i(z)\vartheta_{jk\ell}$$

$$+ \frac{1}{4}\vartheta_{ij}\vartheta_{k\ell}\Big) + \cdots .$$

Now expand in t via $\delta_i = \int_0^t \omega_i(t)\,dt$:

$$= \vartheta(z)^2 + t^2\cdot\sum v_{io}v_{jo}(\vartheta\cdot\vartheta_{ij} - \vartheta_i\vartheta_j) + \frac{t^3}{2}\sum(v_{io}v_{ji}+v_{jo}v_{i1})(\vartheta\cdot\vartheta_{ij} - \vartheta_i\vartheta_j)$$

$$+ t^4\Big[\sum_{i,j,k,\ell=1}^{g} v_{io}v_{jo}v_{ko}v_{\ell o}\Big(\frac{1}{12}\vartheta\cdot\vartheta_{ijk\ell} - \frac{1}{3}\vartheta_i\vartheta_{jk\ell} + \frac{1}{4}\vartheta_{ij}\vartheta_{k\ell}\Big)$$

$$+ \sum_{i,j=1}^{g}\Big(\frac{v_{io}v_{j2}}{6} + \frac{v_{i1}v_{j1}}{4} + \frac{v_{i2}v_{jo}}{6}\Big)(\vartheta\cdot\vartheta_{ij} - \vartheta_i\vartheta_j$$

$$= \vartheta^2 + t^2(\vartheta^2 \cdot D_a^2 \log\vartheta) + t^3(\vartheta^2 \cdot D_a D_a' \log\vartheta) + t^4[\tfrac{1}{12}\vartheta \cdot D_a^4\vartheta - \tfrac{1}{3} D_a\vartheta \cdot D_a^3\vartheta + \tfrac{1}{4}(D_a^2\vartheta)^2$$

$$+ \tfrac{1}{3}(D_a D_a'' \log\vartheta) \cdot \vartheta^2 + \tfrac{1}{4}(D_a'^2 \log\vartheta)\vartheta^2].$$

Now substitute what we have in (3b). Remarkably, the t^i-terms for $i < 4$ cancel and the t^4 terms give:

$$\vartheta^2 \cdot 2c_1 \, D_a^2 \log\vartheta + \tfrac{1}{2} D_a D_a'' \log \vartheta(z) \cdot \vartheta^2$$

$$= (10c_2 - 3c_1^2)\vartheta^2$$

$$+ \tfrac{1}{12}\vartheta \cdot D_a^4\vartheta - \tfrac{1}{3} D_a\vartheta \, D_a^3\vartheta + \tfrac{1}{4}(D_a^2\vartheta)^2$$

$$+ \tfrac{1}{3}(D_a D_a'' \log\vartheta) \cdot \vartheta^2 + \tfrac{1}{4}(D_a'^2 \log\vartheta)\vartheta^2.$$

Use the following lemma:

Lemma. $\frac{1}{2} D^4 \log f + 3[D^2 \log f]^2 = \frac{1}{2}\frac{D^4 f}{f} - 2\frac{Df\, D^3 f}{f^2} + \frac{3}{2}\frac{(D^2 f)^2}{f^2}$.

Proof: Completely straightforward.

We have:

$$2c_1 \, D_a^2 \log\vartheta + \tfrac{1}{6} D_a D_a'' \log\vartheta$$

$$= (10c_2 - 3c_1^2) + \tfrac{1}{4} D_a'^2 \log\vartheta + \tfrac{1}{12} D_a^4 \log\vartheta + \tfrac{1}{2} D_a^2 \log\vartheta .$$

QED

As in §2, these analytic identities have a geometric interpretation in terms of the Kummer variety $\phi(\mathrm{Jac}(X)) \subset \mathbb{P}^{(2^g-1)}$.

Geometric Corollary of (1): Let $|2\Theta|$ define $\phi\colon \mathrm{Jac}(X) \longrightarrow \mathbb{P}^{(2^g-1)}$. Then for all $a,b,c \in \tilde{X}$, then the images under ϕ of

1) the point $\frac{1}{2}\int_a^c \vec{\omega}$

2) the infinitely near point $(\frac{1}{2}\int_a^c \vec{\omega}) + \varepsilon \cdot D_b$

3) the point $\frac{1}{2}\int_{2b}^{a+c} \vec{\omega}$

are collinear, i.e., there is a line in $\mathbb{P}^{(2^g-1)}$ tangent to $\phi(\mathrm{Jac}(x))$ at $\phi(\frac{1}{2}\int_a^c \vec{\omega})$ along the direction D_b and meeting $\phi(\mathrm{Jac}(X))$ at $\phi(\frac{1}{2}\int_{2b}^{a+c} \vec{\omega})$ also.

Proof: This is clearly the limiting form of the Geometric Corollary in §2. Alternatively, we can write (1) as

$$D_b^{(y)}\left[\vartheta(\vec{z}+\vec{y})\cdot\vartheta(\vec{z}-\vec{y})\right]\Big|_{\vec{y}=\frac{1}{2}\int_a^c \vec{\omega}}$$

$$= c_1\vartheta\left(\vec{z}+\frac{1}{2}\int_a^c\right)\vartheta\left(\vec{z}-\frac{1}{2}\int_a^c\right) + c_2\vartheta\left(\vec{z}+\frac{1}{2}\int_{2b}^{a+c}\right)\vartheta\left(\vec{z}-\frac{1}{2}\int_{2b}^{a+c}\right)$$

where

$$D_b^{(y)} = \sum v_{oi}\,\partial/\partial y_i\,, \qquad \text{and} \qquad \omega_i(b) = v_{oi}\,dt\,.$$

Applying the addition theorem and Q as in §2, this gives us

$$D_b\vartheta[{}^0_\eta](\vec{y}, \tfrac{\Omega}{2})\Big|_{\vec{y}=\frac{1}{2}\int_a^c\vec{\omega}} = c_1\vartheta[{}^0_\eta](\tfrac{1}{2}\int_a^c, \tfrac{1}{2}\Omega) + c_2\vartheta[{}^0_\eta](\tfrac{1}{2}\int_{2b}^{a+c}, \tfrac{1}{2}\Omega).$$

where c_1, c_2 are independent of η. QED

When c approaches a, $\phi(\frac{1}{2}\int_a^c)$ approaches $\phi(0)$ which is a singular point of the Kummer Variety. In fact, the local coordinates in $\mathbb{P}^{(2^g-1)}$ at $\phi(0)$ all pull back to even functions on Jac(X). In this situation, elements of $\mathrm{Symm}^2(T_{\mathrm{Jac}(X),0})$ define tangent vectors to $\phi(\mathrm{Jac}(X))$ by the formula:

if (A_{ij}) is a symmetric g×g matrix, let t_A be the vector at $\phi(0)$ given by

$$t_A(f) = \sum_{i,j} \frac{\partial^2}{\partial z_i \partial z_j}(f_\alpha \circ \phi)(0) \cdot A_{ij}$$

for all coordinate functions f_α on $\mathbb{P}^{(2^g-1)}$ near $\phi(0)$.

In particular, if $a,b \in X$, then we get a tangent vector $t_{(a,b)}$ by

$$t_{(a,b)}(f_\alpha) = D_a D_b(f_\alpha \circ \phi)(0) .$$

(This is the case $A_{ij} = \frac{1}{2}(v^{(a)}_{i,o} \cdot v^{(b)}_{j,o} + v^{(a)}_{j,o} \cdot v^{(b)}_{io})$, where $\omega_i(a) = v^{(a)}_{i,o} dt_a$, $\omega_i(b) = v^{(b)}_{i,o}\, dt_b$).

<u>Geometric Corollary of (2):</u> <u>For all</u> $a,b \in \tilde{X}$, <u>the point</u> $\phi(0)$, <u>the vector</u> $t_{(a,b)}$ <u>and the point</u> $\phi(\int_b^a \vec{\omega})$ <u>are collinear</u>.

Proof: To see that this is the correct limiting form of the previous assertion, note that (2) can be rewritten:

$$D_a^{(y)} D_b^{(y)} \vartheta(\vec{z}+\vec{y}) \cdot \vartheta(\vec{z}-\vec{y}) \Big|_{\vec{y}=0}$$

$$= c_1 \vartheta(\vec{z})^2 + c_2 \vartheta(\vec{z} + \int_a^b) \vartheta(\vec{z} - \int_a^b).$$

Applying the addition formula and Q, we get

$$D_a D_b \vartheta[{}^0_\eta](\vec{z}, \tfrac{\Omega}{2}) \Big|_{\vec{z}=0} = c_1 \vartheta[{}^0_\eta](0, \tfrac{\Omega}{2}) + c_2 \vartheta[{}^0_\eta](\int_a^b \vec{\omega}, \tfrac{\Omega}{2}),$$

where c_1, c_2 are independent of η. QED

A different limiting case of (1) is when c approaches b rather than a. Analytically, the constants c_1, c_2 will approach ∞, but geometrically the meaning is that $\phi(\mathrm{Jac}(X))$ will have a point of inflection at $\phi(\frac{1}{2}\int_a^b \vec{\omega})$. This has been used very effectively by Welters and Arbarello-de Conchini in their work on the Schottky problem: cf. Introduction.

Another interpretation of formula (1) shows how the Riemann surface X is intertwined with the function theory of Jac(X). For $a, c \in X$, let $V_{a,c}$ be the vector space of second order theta functions on Jac(X) spanned by the functions

$$\vartheta(z + \tfrac{1}{2}\int_a^c) . \vartheta(z - \tfrac{1}{2}\int_a^c)$$

$$\vartheta(z + \tfrac{1}{2}\int_a^c) . \frac{\partial}{\partial z_i} \vartheta(z - \tfrac{1}{2}\int_a^c) - \vartheta(z - \tfrac{1}{2}\int_a^c) . \frac{\partial}{\partial z_i} \vartheta(z + \tfrac{1}{2}\int_a^c),$$

$$1 \le i \le g.$$

<u>Lemma</u>: $\dim V_{a,c} = g+1$.

<u>Proof</u>: If not, then for some constant c and vector $D = \sum a_i\ \partial/\partial z_i$ there would be an identity:

$$c\,\vartheta\Big(z + \tfrac{1}{2}\int_a^c\Big).\vartheta\Big(z - \tfrac{1}{2}\int_a^c\Big) = \vartheta\Big(z + \tfrac{1}{2}\int_a^c\Big).D\vartheta\Big(z - \tfrac{1}{2}\int_a^c\Big) - \vartheta\Big(z - \tfrac{1}{2}\int_a^c\Big).D\vartheta\Big(z + \tfrac{1}{2}\int_a^c\Big).$$

Let $w = z - \frac{1}{2}\int_a^c$. Then

$$\vartheta(w) = 0 \Longrightarrow \vartheta\Big(w + \int_a^c\Big)\cdot D\vartheta(w) = 0.$$

Since $\vartheta(w + \int_a^c) \neq 0$ for almost all w such that $\vartheta(w) = 0$, this means that $\vartheta(w) = 0 \Longrightarrow D\vartheta(w) = 0$ which we have seen never holds unless $D = 0$. <u>QED</u>

Using $V_{a,c}$ and formula (1), we can recover X as follows:

<u>Corollary</u>:

$$V_{a,c} \cap \begin{pmatrix}\text{locus of decomposable} \\ \text{functions} \\ \vartheta(z+e)\,\vartheta(z-e),\, e \in \mathbb{C}^g\end{pmatrix}$$

$$= \begin{pmatrix}\text{set of functions} \\ \vartheta\Big(z + \frac{1}{2}\int_{2b}^{c+a}\Big)\vartheta\Big(z - \frac{1}{2}\int_{2b}^{c+a}\Big), \\ b \in X\end{pmatrix}$$

$$\cong \text{ cone over } X.$$

Proof: $\supseteq$: This follows from identity (1).

$\subseteq$: Suppose $\vartheta(z+e).\vartheta(z-e) \in V_{a,c}$. Note that if

$\vartheta(z + \frac{1}{2}\int_a^c) = \vartheta(z - \frac{1}{2}\int_a^c) = 0$ then all functions in $V_{a,c}$ vanish.

Therefore

$$\vartheta(z + \tfrac{1}{2}\int_a^c) = \vartheta(z - \tfrac{1}{2}\int_a^c) = 0 \Longrightarrow \vartheta(z+e) = 0 \text{ or } \vartheta(z-e) = 0.$$

Substituting $z + \frac{1}{2}\int_a^c$ for z and $e - \frac{1}{2}\int_a^c$ for e, this says:

$$\vartheta(z + \int_a^c) = \vartheta(z) = 0 \Longrightarrow \vartheta(z + e) = 0 \text{ or } \vartheta(z + \int_a^c - e) = 0.$$

We will show that if this holds, then $e = \int_b^c \vec{\omega}$ or $e = \int_a^b \vec{\omega}$ which,

substituting back, is what we want.

Our hypothesis can be written

$$(*) \qquad \Theta \cap \Theta_{\left(\int_a^c \vec{\omega}\right)} = \Theta_e \cup \Theta_{\left(\int_a^c \vec{\omega} - e\right)}$$

where Θ_f is Θ translated by f. Next, use Riemann's theorem to express this in terms of divisors. To fix notation, let $\vec{z} \in \mathbb{C}^g$ define the divisor class D(z) by

$$\int_{D(z)} \vec{\omega} = \vec{z},$$

and let δ be the divisor class of degree g-1 such that

$$z \in \Theta \Longleftrightarrow |D(z)+\delta| \neq \emptyset.$$

Let $D_z = D(z)+\delta$. So, $z \in \Theta \cap \Theta_{\left(\int_a^c \omega\right)} \Longleftrightarrow |D_z| \neq \emptyset$ and $|D_z + c - a| \neq \emptyset$.

Let W = set of divisors D_z such that $z \in \Theta \cap \Theta_{\left(\int_a^c \omega\right)}$. Clearly W contains the subset $W_a = \{\text{divisors } D_o + a\colon\ D_o = \sum_{i=1}^{g-2} Q_i\}$. Our hypothesis (*) tells us that:

$$D_z \in W \Longrightarrow \text{either } |D_z+D(e)| \neq \emptyset \quad \text{or} \quad |D_z + c - a - D(e)| \neq \emptyset.$$

Since W_a is an irreducible set, it must lie entirely in one of these two sets:

$$D = D_0+a \in W \Longrightarrow \text{either } |D_0+a + D(e)| \neq \emptyset \quad \forall\, D_0 = \sum_{i=1}^{g-2} Q_i$$
$$\text{or } |D_0+c - D(e)| \neq \emptyset \quad \forall\, D_0 = \sum_{i=1}^{g-2} Q_i.$$

The following lemma then finishes this proof.

<u>Lemma</u>: <u>If</u> $D(e)$ <u>is a divisor of degree zero such that for all</u> $D_o = \sum_{i=1}^{g-2} Q_i$,

$$|D_0 + a + D(e)| \neq \emptyset\ ,$$

<u>then</u> $D(e) \sim b-a$ <u>for some</u> $b \in X$.

<u>Proof</u>. Left to the reader.

So, we have used formula (1) to construct the cone over X, and hence X. We can ask whether we can also use formula (2) to construct X. As a possible approach, start out as above, and let

V_o be the vector space spanned by the functions

$$\vartheta(z)^2, \qquad \vartheta(z)\cdot\frac{\partial^2\vartheta}{\partial z_i\partial z_j} - \frac{\partial\vartheta}{\partial z_i}\cdot\frac{\partial\vartheta}{\partial z_j}, \qquad 1 \le i \le j \le g.$$

As above: a) $V_0 \subset$ vector space of second order ϑ-functions

b) $\dim V_0 \le 1 + \frac{g(g+1)}{2}$.

Consider $V_0 \cap \begin{pmatrix}\text{decomposition functions} \\ \vartheta(z+a)\cdot\vartheta(z-a).\end{pmatrix}$

Formula 5.2 tells us that this <u>contains</u> the set:

$$\{\vartheta(z + \int_a^b)\ \vartheta(z - \int_a^b) \quad \text{some } a,b \in X\}$$

which is isomorphic to a cone over $\mathrm{Symm}^2 X$.

<u>Question</u> 1. Are these two spaces equal?

This would follow, as above, from the following question:

<u>Question</u> 2. If D is a divisor class of degree 0 on X such that for all divisors E of degree g-1 for which $|E|$ is a pencil, then either $|D+E| \neq \emptyset$ or $|D-E| \neq \emptyset$, then does it follow that $D \sim a-b$ for some $a,b \in X$?

§4. Applications to solutions of differential equations

The corollaries of Fay's trisecant identity can be used to construct special solutions to many equations occurring in Mathematical Physics. In this section we will consider the following equations:

1) Sine-Gordan: $u_{tt} - u_{xx} = \sin u.$

2) Korteweg-de Vries (K-dV): $u_t + u_{xxx} + u \cdot u_x = 0.$

3) Kadomtsev-Petviashvili (K-P): $u_{yy} + (u_t + u_{xxx} + u \cdot u_x)_x = 0.$

Many other equations also have solutions constructed via theta functions, such as

4) Non-linear Schrödinger: $iu_t = u_{xx} \pm u \cdot |u|^2.$

5) Massive Thirring model: $i\,u_x = v(1 + u\bar{v})$
$i\,v_y = u(1 + v\bar{u}),$

but we will not consider these here (for the non-linear Schrödinger equation, see the PhD thesis of E. Previato, Harvard, 1983).

We will give some solutions in terms of ϑ-functions to the first three equations. In the last section, we will indicate how one uses the generalized Jacobian to relate our solutions to the famous "soliton" solutions to the K-dV equation.

The easiest solutions to obtain are some for the K-P equation.

Corollary. For all $a \in X$, $12\, D_a^2 \log \vartheta(\vec{z}_0 + x\vec{v}_0 + \sqrt{3}\, y\vec{v}_1 - 2t\vec{v}_2) + 2c_1$ satisfies K-P, where:

c_1 is the constant appearing in formula (3), §3.

$\vec{v}_i = (v_{ij}, \dots, v_{gj})$

$\omega_i = \sum v_{ij} \frac{t^j}{j!}\, dt$, (t a local coordinate near a)

<u>Proof</u>. Take D_a^2 of formula (3) and set $u(z) = D_a^2 \log \vartheta(z)$ to get:

$$D_a^4 u(z) + 12 D_a^2 u(z) \cdot u(z) + 12 (D_a u(z))^2 + 2c_1 D_a^2 u(z) - 2 D_a D_a'' u(z) + 3 (D_a')^2 u(z) = 0.$$

Let $v = 12u + 2c_1 = 12 D_a^2 \log \vartheta(z) + 2c_1$; then:

$$3 D_a'^2 v(z) + D_a (D_a^3 v(z) + v(z) \cdot D_a v(z) - 2 D_a'' v(z)) = 0.$$

Finally, note that by definition,

$$D_a u(z) = \frac{\partial}{\partial x} u(\vec{z} + x \cdot \vec{v}_0),$$

$$D_a' u(z) = \frac{\partial}{\partial y} u(\vec{z} + y \cdot \vec{v}_1),$$

$$D_a'' u(z) = \frac{\partial}{\partial t} u(\vec{z} + t \cdot \vec{v}_2).$$

Thus

$$v(\vec{z}_0 + x\vec{v}_0 + \sqrt{3}\, y\, \vec{v}_1 - 2t\vec{v}_2)$$

solves K-P, as wanted. <u>QED</u>

In order to find solutions to KdV and Sine-Gordan, we need to consider hyperelliptic curves. Let X be hyperelliptic, $\pi: X \longrightarrow \mathbb{P}^1$ the double cover, and let $i: X \longrightarrow X$ be the involution.

Let $a \in X$ be a branch point of π and let t be a local coordinate about a such that the hyperelliptic involution i is just $t \longmapsto -t$. But $i^*\omega_j = -\omega_j$ (see Ch. (IIIa, §2) so if $\omega_j = v_j(t)dt$, $v_j(t)dt + v_j(-t)d(-t) = 0$. Thus v_j is an even function of t, hence $v_{j1} = 0$ and $D_a' = 0$.

Corollary. $12\, D_a^2 \log \vartheta(\vec{z}_0 + x\vec{v}_0 + t\vec{v}_2)) + 2c_1$ satisfies KdV

where:

$c_1, \vec{v}_0, \vec{v}_2$ are as in the previous corollary

and $a \in X$, X hyperelliptic, and the local coordinate t at a satisfies $i^*t = -t$.

Proof. Take D_a of formula (3), and use the above fact that $D_a' = 0$ to get the result.

Next we would like to tackle Sine-Gordan. Recall from Ch. IIIa that if $a,b \in X$ are branch points, then $\int_a^b \vec{\omega} \in \frac{1}{2} L_\Omega$, i.e., if a,b are branch points, $\int_a^b \vec{\omega} = \frac{1}{2}(n+\Omega m)$ for some $n,m \in \mathbb{Z}^g$,

To solve Sine-Gordan: Let X be hyperelliptic $a,b \in X$ branch points. Start with Formula (2). Substitute $z \longrightarrow z + \int_a^b$ and subtract the original formula:

$$D_a D_b \log \frac{\vartheta(z+\int_a^b)}{\vartheta(z)} = c_2 \frac{\vartheta(z+2\int_a^b)\vartheta(z)}{\vartheta(z+\int_a^b)^2} - \frac{\vartheta(z+\int_a^b)\vartheta(z+\int_a^b - 2\int_a^b)}{\vartheta(z)^2} .$$

Let $\int_a^b \vec{\omega} = \frac{1}{2}(n+\Omega m)$ and get, using the functional equation for ϑ:

$$D_a D_b \log \frac{\vartheta(z+\int_a^b)}{\vartheta(z)} = c_2 \Bigg[e^{-\pi i {}^t m\Omega m} e^{-2\pi i\, {}^t m z} \cdot \frac{\vartheta(z)^2}{\vartheta(z+\int_a^b)^2} - e^{\pi i\, {}^t m\cdot n} e^{2\pi i\, {}^t m z} \cdot \frac{\vartheta(z+\int_a^b)^2}{\vartheta(z)^2} \Bigg]$$

$$\text{Let } u(z) = 2i \log \frac{\vartheta(z+\int_a^b)}{\vartheta(z)} - 2\pi\, {}^t m(z + \frac{1}{2}\int_a^b); \text{ then}$$

$$D_a D_b\, u(z) = c_2'\left[\frac{e^{iu(z)} - e^{-iu(z)}}{2i}\right],$$

where

$$c_2' = -4c_2\, e^{-\pi i \cdot \frac{1}{2}\, {}^t m\Omega m + \frac{1}{2}\pi i\, {}^t mn} .$$

So $u(z)$ satisfies $D_a D_b u(z) = c_2' \cdot \sin u(z)$. Thus for any $\vec{z}_0$, the function

$$v(x,t) = u(\vec{z}_0 + x(\frac{\vec{a}-\vec{b}}{2}) + t(\frac{\vec{a}+\vec{b}}{2}))$$

satisfies

$$\frac{\partial^2}{\partial t^2} v(x,t) - \frac{\partial^2}{\partial x^2} v(x,t) = c_2' \cdot \sin v(x,t),$$

where $\vec{a},\vec{b}$ are proportional to $(w_1(a),\ldots,w_g(a))$ and $(w_1(b),\ldots,w_g(b))$ respectively. We pass over the interesting question of when v and c_2' are real and what these solutions "look like".

§5. The Generalized Jacobian of a Singular Curve and its Theta Function

In this section we will define and describe the generalized Jacobian of the simplest singular curves: the curves obtained by identifying 2g points of $\mathbb{P}^1$ in pairs. We will then determine their theta functions and theta divisors. Finally, we will apply this theory to understand analytically and geometrically the limits of the solutions to the KdV equation that were discussed in the previous section, when the hyperelliptic curve becomes singular of the above form.

Let C be a singular curve of genus g, and let $S = \mathrm{Sing}(C)$. Suppose the singularities of C are only nodes $p_1,\ldots,p_g$ and that C has normalization $\pi\colon \mathbb{P}^1 \longrightarrow C$. If $\pi^{-1}(p_i) = \{b_i,c_i\}$, $i = 1\ldots g$, this means that C is just $\mathbb{P}^1$ with the g pairs of points $\{b_i,c_i\}$ identified. We assume $\infty \neq b_i,c_i$ $\forall i$. Now, in general we define

$$\mathrm{Pic}\ C = \left\{\begin{array}{l} \text{group of divisors } D = \Sigma n_i x_i,\ \ x_i \in C-S \\ \text{mod: } D \sim 0 \text{ if } D = (f) \text{ for} \\ \text{some } f \in \mathbb{C}(C),\ f \text{ continuous and} \\ \text{finite, nonzero at each } p_i \end{array}\right\}$$

In our case we can pull back to $\mathbb{P}^1$ and we get

$$\mathrm{Pic}\ C = \left\{\begin{array}{l} \text{group of divisors } D = \Sigma n_i x_i,\ \ x_i \in \mathbb{P}^1 - \pi^{-1}(S) \\ \text{mod: } D \sim 0 \text{ if } \ D = (f),\ \ f \in \mathbb{C}(\mathbb{P}^1) \\ \text{and } f(b_i) = f(c_i) \text{ for all } i = 1\ldots g \end{array}\right\}$$

We define Jac(C) to be the piece $\mathrm{Pic}^o(C)$ of Pic(C) corresponding to divisors $n_i x_i$ of degree 0, i.e., $\Sigma n_i = 0$. The structure of

Jac(C) is easy to work out : start with D of degree 0. As a divisor on $\mathbb{P}^1$, it equals the divisor of zeroes and poles of some rational function f. The ratios $f(b_i)/f(c_i)$ represent the obstruction to D being zero in Pic(C). It is easy to verify that they set up an isomorphism of groups:

$$\mathrm{Jac}(C) \xrightarrow{\ \sim\ } (\mathbb{C}^*)^g$$

$$D \longmapsto \left(\frac{f(b_1)}{f(c_1)}, \ldots, \frac{f(b_g)}{f(c_g)}\right).$$

As in chapter IIIa, we can add to any divisor D the divisor $x-x_0$ and get a family of divisors $D+x-x_0$ depending on a point x near x_0. Letting x approach x_0 this gives a tangent vector to Jac C near D, and as D varies, an invariant vector field D_{x_0} on Jac C. For later use we can work out this vector field in terms of coordinates $\lambda_1,\ldots,\lambda_g$ on $(\mathbb{C}^*)^g$:

If $D = (f(t))$, then $D+x-x_0 = (f(t)\cdot \frac{t-x}{t-x_0})$; hence the coordinates of $D+x-x_0$ in Jac C are

$$\lambda_i = \frac{f(b_i)\cdot \dfrac{b_i-x}{b_i-x_0}}{f(c_i)\cdot \dfrac{c_i-x}{c_i-x_0}} .$$

Then

$$\left.\frac{\partial\lambda_i}{\partial x}\right|_{x=x_0} = \left.\frac{f(b_i)}{f(c_i)}\cdot\frac{c_i-x_0}{b_i-x_0}\cdot\frac{b_i-c_i}{(c_i-x)^2}\right|_{x=x_0}$$

$$= \lambda_i\Big|_{x=x_0} \frac{b_i-c_i}{(b_i-x_0)(c_i-x_0)} .$$

Thus the vector field D_{x_0} is given by

$$D_{x_0} = \sum_{i=1}^{g} \frac{b_i - c_i}{(b_i - x_0)(c_i - x_0)} \lambda_i \frac{\partial}{\partial \lambda_i} .$$

Now, Jac C is not compact: we want to construct a natural compactification of it. N.B. This will no longer be a group however! It is clear what we need to do to compactify: we need to allow the support of our divisors to approach the singular points. But considering divisors $\Sigma n_i x_i$, arbitrary $x_i \in C$ does not work very well. We need to encode the "multiplicity" of the singular point in a more subtle way. This is done as follows. In general let

$$\overline{\text{Pic } C} = \left\{ \begin{array}{l} \underline{\text{set}} \text{ of coherent } \mathcal{O}_C\text{-module } \mathcal{F} \subset \mathbb{C}(C) \\ \text{up to isomorphism} \end{array} \right\}$$

Translating this to more down-to-earth language, this becomes

set of all divisors $D = \sum_{i=1}^{k} n_i x_i$ along with finitely generated $\mathcal{O}_{C,x_i}$-modules $M_{x_i} \subset \mathbb{C}(C)$, ($x_i \in C$) are arbitrary, where

if x_i is not singular, M_{x_i} is simply $t^{-n_i}\mathcal{O}_{C,x_i}$, t a local coordinate near x_i,

and if x_i is singular, n_i is determined via:

$$n_i = \dim \frac{M_{x_i}}{M_{x_i} \cap \mathcal{O}_{C,x_i}} - \dim \frac{\mathcal{O}_{C,x_i}}{M_{x_i} \cap \mathcal{O}_{C,x_i}} .$$

By convention, $M_x = \mathcal{O}_{C,x}$ if $x \notin \{x_1, \cdots, x_k\}$.

mod: $D \sim D'$ if $\exists f \in \mathbb{C}(C)$ such that $M_x = f \cdot M'_x$, $\forall x \in C$.

The modules M_{x_i} can be thought of as a refined way of measuring the multiplicity n_i at the singular points: we will call them the multiplicity modules. $\overline{\text{Pic } C}$ always has a natural structure of projective variety but let's just think of it as a set.

In our case of g nodes, we know exactly what the M_{x_i}'s must look like:

Lemma. If $p \in C$ is an ordinary double point obtained by glueing two points b,c in a smooth curve $\tilde{C}$, then for all $M \subset \mathbb{C}(C)$ which are finitely generated $\mathcal{O}_{P,C}$-modules, either:

a) $M = f \cdot \mathcal{O}_{P,C}$ for some $f \in \mathbb{C}(C)$

or

b) $M = f \cdot \tilde{\mathcal{O}}_{P,C}$ for some $f \in \mathbb{C}(C)$, where $\tilde{\mathcal{O}}_{P,C} = \mathcal{O}_{b,\tilde{C}} \cap \mathcal{O}_{c,\tilde{C}} =$ normalization of $\mathcal{O}_{P,C}$.

Proof. Let $M_{k,\ell}$ = module of functions f such that $\text{ord}_a f \geq k$, $\text{ord}_b f \geq \ell$. Take k,ℓ the largest integers so that $M \subset M_{k,\ell}$. Then almost all functions $f \in M$ satisfy $\text{ord}_a f = k$ and $\text{ord}_b f = \ell$. So choose such an $f \in M$. We have

$$f \cdot \mathcal{O}_{P,C} \subset M \subset M_{k,\ell} .$$

But now $M_{k,\ell} = f \cdot M_{0,0}$ and $M_{0,0}$ is just $\tilde{\mathcal{O}}_{P,C}$. Moreover, $\mathcal{O}_{P,C}$ is the subspace of $\tilde{\mathcal{O}}_{P,C}$ defined as $\{g \mid g(b)=g(c)\}$ so it has codimension 1 in $\tilde{\mathcal{O}}_{P,C}$. Therefore

$$\dim M_{k,\ell}/f \cdot \mathcal{O}_{P,C} = \dim f \cdot \tilde{\mathcal{O}}_{P,C}/f \cdot \mathcal{O}_{P,C} = \dim \tilde{\mathcal{O}}_{P,C}/\mathcal{O}_{P,C} = 1.$$

So either $M = f \cdot \mathcal{O}_{P,C}$ or $M = M_{k,\ell} = f \cdot \tilde{\mathcal{O}}_{P,C}$, as wanted. QED

From this lemma, we get immediately:

Corollary. For any subset $T \subset \{P_1,\ldots,P_g\}$, let $C_T = [C$ with P_i separated into b_i and i_1 for $i \in T] = [\mathbb{P}^1$ with b_i, c_i identified for $i \notin T]$. Then, as a set:

$$\overline{\mathrm{Pic}\, C} = \coprod_T \mathrm{Pic}(C_T) \qquad (\coprod = \text{disjoint union})$$

Proof: In fact, divide up all divisors $D = \{\Sigma n_i x_i, M_i\}$ according to whether their multiplicity modules are isomorphic to $\mathcal{O}_{P_i,C}$ or $\tilde{\mathcal{O}}_{P_i,C} = \mathcal{O}_{b_i,\mathbb{P}^1} \cap \mathcal{O}_{c_i,\mathbb{P}^1}$ at each singular point. For each subset $T \subset \{P_1,\cdots,P_g\}$, let $\overline{\mathrm{Pic}(C)}^{(T)}$ be the set of D whose multplicity module is $\tilde{\mathcal{O}}_{P_iC}$ exactly for $P_i \in T$. We claim:

$$\overline{\mathrm{Pic}(C)}^{(T)} \cong \mathrm{Pic}(C_T).$$

In fact, if $D \in \overline{\mathrm{Pic}(C)}^{(T)}$, then when $P_i \notin T$, P_i singular, there exists an f_i such that $M_{P_i} = f_i \mathcal{O}_{P_i,C}$. It's not hard to see that one can choose a single rational function f such that this holds for all such P_i.

Let D' be defined by the multiplicity modules $f^{-1} \cdot M_P$. It defines a divisor on C_T with "trivial" multiplicity $\mathcal{O}_{P_i,C}$ at all the singularities of C_T. Two such are equivalent in $\mathrm{Pic}(C_T)$ if and only if they are equivalent in $\overline{\mathrm{Pic}(C)}$ because the condition $f(b_\ell) = f(c_\ell)$ in the definition of equality in $\mathrm{Pic}(C_T)$ is the same as the condition $f \cdot \mathcal{O}_{P_\ell,C} = \mathcal{O}_{P_\ell,C}$ included in the definition of equality in $\overline{\mathrm{Pic}(C)}$.

Actually, we can be much more explicit, and make the degree 0 component $\overline{\text{Jac } C}$ into a compact analytic space as follows:

<u>Theorem</u> $\overline{\text{Jac}(C)} \cong (\mathbb{P}^1)^g/\sim$, <u>with equivalence relation</u>

$$(\omega_{k1}\lambda_1, \ldots, \underset{\substack{\uparrow \\ k^{th}\text{ spot}}}{\infty}, \ldots, \omega_{kg}\lambda_g) \sim (\lambda_1, \ldots, \underset{\substack{\uparrow \\ k^{th}\text{ spot}}}{0}, \ldots, \lambda_g), \qquad \text{for all } k.$$

<u>where</u> $\omega_{ij} = \dfrac{(b_i-b_j)(c_i-c_j)}{(b_i-c_j)(c_i-b_j)}$.

<u>Sketch of proof</u>: Fix some $n \geq g$ and let

$S = \left(\underline{\text{unordered sets}}\ (x_1,\ldots,x_n) : x_i \in \mathbb{P}^1;\ \text{for each } i,\ \exists \text{ at most one } j \text{ s.t. } x_j \in \{b_i, c_i\}\right)$.

Define two maps

$$(\mathbb{P}^1)^g \xleftarrow{\ \pi_1\ } S \xrightarrow{\ \pi_2\ } \overline{\text{Jac}(C)}$$

by $\pi_1(x_1,\cdots,x_n) = \left(\prod_{i=1}^{n} \dfrac{b_1-x_i}{c_1-x_i}, \ldots, \prod_{i=1}^{n} \dfrac{b_g-x_i}{c_g-x_i}\right)$, and

$\pi_2(x_1,\ldots,x_n) = \left(\text{the divisor } x_1+\cdots+x_n - n\cdot\infty\right)$, where if $x_i = b_j$ or c_j, the multiplicity module is $\mathfrak{m}_{p_j}$ (the maximal ideal of functions zero at p_i).

The following things are not hard to prove:

a) if n is sufficiently large, e.g., 2g, then π_1, π_2 are surjective

b) π_2 is constant on the fibres of π_1 so that there is a unique map $\varphi: (\mathbb{P}^1)^g \longrightarrow \overline{\text{Jac}(C)}$ satisfying $\varphi \circ \pi_1 = \pi_2$.

c) φ is independent of n and defines an isomorphism of $(\mathbb{P}^1)^g/\sim$ with $\overline{\mathrm{Jac}(C)}$.

d) φ restricted to $(\mathbb{C}^*)^g$ is the isomorphism

$$(\mathbb{C}^*)^g \xrightarrow{\sim} \mathrm{Jac}(C)$$

defined above.

e) More generally, if $T \subset \{1,\cdots,g\}$ is any subset, $h = g - \#T$ and $\varepsilon: T \longrightarrow \{0,\infty\}$ any function, then φ restricted to

$$\prod_{i \in T} \{\varepsilon(i)\} \times \prod_{i \notin T} \mathbb{C}^* \subset (\mathbb{P}^1)^g$$

is the same isomorphism of $(\mathbb{C}^*)^h \xrightarrow{\sim} \mathrm{Jac}(C^T) \subset \overline{\mathrm{Jac}(C)}$ up to multiplication by a constant in $(\mathbb{C}^*)^h$.

The idea of the crucial step b is this:

Say $\pi_1(x_1,\cdots,x_n) = \pi_1(y_1,\cdots,y_n)$, <u>and</u> $x_i, y_i \in \mathbb{P}^1 - \bigcup_k \{b_k, c_k\}$. Let

$$f(t) = \prod_{\substack{1 \le i \le n \\ x_i \neq \infty}} (t - x_i) \Big/ \prod_{\substack{1 \le i \le n \\ y_i \neq \infty}} (t - y_i)$$

then the hypothesis says that

$$f(b_k) = f(c_k), \qquad \text{all } k$$

hence

$$\left(\sum x_i - n\cdot\infty\right) \sim \left(\sum y_i - n\cdot\infty\right) \quad \text{in} \quad \mathrm{Pic}(C).$$

In Step C, the ω's come in because for any $x_2,\cdots,x_n \in \mathbb{P}^1 - \bigcup_k \{b_k, c_k\}$, we have $\pi_2(b_k,x_2,\cdots,x_n) = \pi_2(c_k,x_2,\cdots,x_n)$, and

$$\pi_1(b_k,x_2,\cdots,x_n) = \left(\frac{b_k-b_1}{b_k-c_1}\cdot\prod_{i=2}^{n}\frac{x_i-b_1}{x_i-c_1},\cdots,0,\cdots,\frac{b_k-b_g}{b_k-c_g}\cdot\prod_{i=2}^{n}\frac{x_i-b_g}{x_i-c_g}\right)$$

$$\pi_1(c_k,x_2,\cdots,x_n) = \left(\frac{c_k-b_1}{c_k-c_1}\cdot\prod_{i=2}^{n}\frac{x_i-b_1}{x_i-c_1},\cdots,\infty,\cdots,\frac{c_k-b_g}{c_k-c_g}\cdot\prod_{i=2}^{n}\frac{x_i-b_g}{x_i-c_g}\right)$$

ratio ω_{k1} k^{th} spot   ratio ω_{kg}

The details of the proof are not central to the exposition and are omitted.

Several points in this proof are useful below. Firstly, note that $\overline{\text{Jac } C}$ has one "most singular" point at infinity, namely the point corresponding to $(\lambda_1,\ldots,\lambda_g)$ where all λ_i are either 0 or ∞. We will call this P_∞. Secondly, the map π_1 enables us to construct an analog of Θ for Jac C. To do this, let's calculate $\dim \pi_1^{-1}(\lambda_1,\cdots,\lambda_g)$.

Let $\pi_1(x_1,\ldots,x_g) = (\lambda_1,\ldots,\lambda_g)$. Up to an undetermined constant, let $\varphi(t) = c \cdot \Pi(t-x_i)$, where if $x_i = \infty$ that term is omitted. So $\deg(\varphi) \leq g$. Write $\varphi(t) = \sum_{i=0}^{g}\varphi_i t^i$. The φ_i depend on $x_1,\cdots,x_g$ and determine $\{x_1,\cdots,x_g$ uniquely up to permutation. Now,

$$\frac{\phi(b_k)}{\phi(c_k)} = \lambda_k \qquad \text{for} \quad k = 1\ldots g .$$

So

$$\sum_{i=0}^{g} \phi_i(\lambda_k c_k^i - b_k^i) = 0 \qquad \text{for} \quad k = 1\ldots g.$$

$\pi_1^{-1}(\lambda_1,\cdots,\lambda_g)$ is given by the set of solutions in $[\phi_i]$ of these equations so

$$\dim \pi_1^{-1}(\lambda_1,\ldots,\lambda_g) = g - \operatorname{rank}_{\substack{i=0\ldots g\\ k=1\ldots g}}(\lambda_k c_k^i - b_k^i)$$

In particular, π_1 is generically 1-1.

Next, let us determine the analog of the theta divisor Θ using the above. We want equations for the locus where the divisor $\sum_{i=1}^{g} x_i - \infty$ is effective. From the discussion above, this is exactly when $\deg \phi \le g-1$, i.e., $\phi_g = 0$. Over a given point $(\lambda_1,\cdots,\lambda_g)$, there is such a ϕ if and only if:

$$\det \begin{vmatrix} 1-\lambda_1 & \cdots & 1-\lambda_g \\ b_1-\lambda_1 c_1 & \cdots & b_g-\lambda_g c_g \\ \vdots & & \vdots \\ b_1^{g-1} - \lambda_1 c_1^{g-1} & \cdots & b_g^{g-1}-\lambda_g c_g^{g-1} \end{vmatrix} = 0$$

This determinant is the analog of ϑ and its zeroes, as a subset of $(\mathbb{P}^1)^g/\sim$ or via φ as a subset of $\overline{\mathrm{Pic}(C)}$, are the analog of Θ.

We shall call this function $\tau_C(\lambda_1,\cdots,\lambda_g)$. τ_C has a useful expansion. First recall the Vandermonde determinant

$$\det \begin{vmatrix} 1 & 1 & \cdots & 1 \\ a_1 & a_2 & \cdots & a_g \\ a_1^2 & a_2^2 & \cdots & a_g^2 \\ \vdots & & & \\ a_1^{g-1} & a_2^{g-1} & \cdots & a_3^{g-1} \end{vmatrix} = \prod_{i>j} (a_i-a_j)$$

In the above determinant, this enables us to work out the coefficient of the $\prod_{i\in S} \lambda_i$ term:

$$\prod_{\substack{i>j\\ i,j\in S}} (c_i-c_j) \cdot \prod_{\substack{i>j\\ i,j\notin S}} (b_i-b_j) \cdot \prod_{\substack{i\notin S\\ j\in S}} (b_i-c_j) \cdot (-1)^{\#S} \cdot a(S)$$

where $a(S)$ = the sign of the permutation changing 1..g to $(S,\{1..g\}-S)$ and preserving the order of each set (e.g., $a(\{1,3\})=-1$). τ_C, therefore, can be expanded:

$$(*) \qquad \tau_C = \prod_{i<j} (b_i-b_j)\cdot \sum_{S\subset\{1..g\}} (-1)^{\#S} \cdot \prod_{i\in S} \Big(\lambda_i \prod_{j\neq i} \frac{c_i-b_j}{b_i-b_j}\Big) \cdot \prod_{\substack{i<j\\ i,j\in S}} \omega_{ij} = 0.$$

Note that the worst boundary point, $P_\infty = (0,\cdots,0)$ is <u>not</u> on Θ, and correspondingly, $\det(0,\ldots,0) \neq 0$.

I claim that this determinant is also a limit of theta functions of our non-singular curves C. Formally, we can see a link as follows:

Let $\Omega_{ij}(t)$ be a family of period matrices in which

$$\operatorname{Im} \Omega_{ii}(t) \longrightarrow \infty \quad \text{as} \quad t \longrightarrow 0, \qquad 1 \le i \le g,$$

and

$$\Omega_{ij}(t) \quad \text{are continuous for } |t| < \varepsilon, \text{ if } \quad i \neq j.$$

Then consider $\vartheta(z, \Omega(t))$. The limit of this function as $t \longrightarrow 0$ will be just 1. A better thing to do is to translate the functions by a vector depending on t first:

Let $\delta\Omega(t)$ = diagonal of $\Omega(t)$; then

$$\vartheta\left(z - \frac{\delta\Omega(t)}{2}, \Omega(t)\right) = \sum_{m \in \mathbb{Z}^g} e^{\pi i\, {}^t m \Omega m + 2\pi i\, {}^t m (z - \frac{\delta\Omega(t)}{2})}$$

$$= \sum_{m \in \mathbb{Z}^g} \prod_{i=1}^{g} e^{\pi i (m_i^2 - m_i)\Omega_{ii}(t)} \cdot \prod_{i<j} e^{2\pi i\, m_i m_j \Omega_{ij}(t)} \cdot e^{2\pi i\, {}^t m z}.$$

As $t \longrightarrow 0$, this function approaches:

$$\sum_{\substack{\vec{m}=(m_1,\cdots,m_g) \\ m_i = 0 \text{ or } 1}} \prod_{i<j} e^{2\pi i \cdot m_i m_j \Omega_{ij}(0)}\, e^{2\pi i\, {}^t m z}$$

$$= \sum_{S \subset \{1..g\}} \prod_{\substack{i<j \\ i,j \in S}} e^{2\pi i \Omega_{ij}(0)} \cdot \prod_{i \in S} \cdot e^{2\pi i z_i}.$$

Now if

$$e^{2\pi i\, \Omega_{ij}(0)} = \frac{(b_i - b_j)(c_i - c_j)}{(b_i - c_j)(c_i - b_j)} = \omega_{ij},$$

and

$$e^{2\pi i z_i} = -\lambda_i \cdot \prod_{j\neq i} \frac{c_i-b_j}{b_i-b_j}$$

it equals τ_C up to a constant. In fact, if C_t is a family of smooth curves of genus g "degenerating" to C, it can be shown that its period matrix behaves exactly like this. Correspondingly, in the lattice $L_{\Omega(t)} = \mathbb{Z}^g + \Omega(t)\mathbb{Z}^g$, the B-periods $\Omega(t)\mathbb{Z}^g$ go to infinity, but the A-periods $\mathbb{Z}^g$ remain finite. Thus $X_{\Omega(t)} = \mathbb{C}^g/\mathbb{Z}^g+\Omega(t)\mathbb{Z}^g$ tends to $\mathbb{C}^g/\mathbb{Z}^g$, which is just $(\mathbb{C}^*)^g$ with coordinates $e^{2\pi i z_i}$. We do not want to describe this in detail, referring the reader to Fay, op. cit., Ch. 3.

In the limit, is there anything left of the quasi-periodicity of ϑ with respect to its B-periods? At first it would seem not but there is, in fact, something. In fact, the three methods by which we formed from ϑ meromorphic functions on X_Ω now give us rational functions on the compactification $\overline{\text{Jac } C}$ which are continuous maps

$$\overline{\text{Jac } C} - (\text{codim 2 set of indeterminacy}) \longrightarrow \mathbb{P}^1 .$$

The point is that the induced rational maps

$$(\mathbb{P}^1)^g - (\text{codim. 2 set}) \longrightarrow \mathbb{P}^1$$

are compatible with the equivalence relation $\sim$ of the above theorem.

Let's check this for the second logarithmic derivative with respect to the invariant vector fields $\lambda_i\, \partial/\partial\lambda_i$, $\lambda_j\, \partial/\partial\lambda_j$, i.e.,

$$\lambda_i \frac{\partial}{\partial\lambda_i} \cdot \lambda_j \frac{\partial}{\partial\lambda_j} (\log \tau_C(\lambda_1,\cdots,\lambda_g)).$$

Note that this is the analog of $\frac{\partial}{\partial z_i}\frac{\partial}{\partial z_j}\log\vartheta(z)$. Let $\lambda_i' = \lambda_i \prod_{j\neq i}\frac{c_i-b_j}{b_i-b_j}$. If $k \neq i,j$, then

$$\lim_{\lambda_k\to\infty} \lambda_i \frac{\partial}{\partial\lambda_i}\cdot\lambda_j\frac{\partial}{\partial\lambda_j}(\log\tau_C(\lambda_1,\ldots,\lambda_g))$$

$$= \lim_{\lambda_k\to\infty}\lambda_i\frac{\partial}{\partial\lambda_i}\cdot\ \lambda_j\frac{\partial}{\partial\lambda_j}[\log\lambda_k' + \log\sum_S(-1)^{\#S}\prod_{\substack{i<j\\ i,j\in S}}\omega_{ij}\cdot\prod_{\substack{i\in S\\ i\neq k}}\lambda_i'\cdot\begin{Bmatrix}1 & \text{if } k\in S\\ \lambda_k'^{-1} & \text{if } k\notin S\end{Bmatrix}$$

$$= \lambda_i\frac{\partial}{\partial\lambda_i}\cdot\lambda_j\frac{\partial}{\partial\lambda_j}[\log\sum_{\substack{S\text{ with}\\ k\in S}}(-1)^{\#S}\cdot\prod_{\substack{i<j\\ i,j\in S-k}}\omega_{ij}\cdot\prod_{i\in S-k}\omega_{ik}\cdot\prod_{i\in S-k}\lambda_i']$$

$$= \lambda_i\frac{\partial}{\partial\lambda_i}\cdot\lambda_j\frac{\partial}{\partial\lambda_j}[\log\sum_{\substack{S\text{ with}\\ k\notin S}}(-1)^{\#S+1}\prod_{\substack{i<j\\ i,j\in S}}\omega_{ij}\cdot\prod_{i\in S}\omega_{ik}\lambda_i']$$

$$= \lim_{\lambda_k\to 0}\lambda_i\frac{\partial}{\partial\lambda_i}\cdot\lambda_j\frac{\partial}{\partial\lambda_j}[\log\tau_C(\omega_{1k}\lambda_1,\ldots,\lambda_k,\ldots,\omega_{gk}\lambda_g)]\ .$$

Now let's apply this to give solutions of KdV. We want C to be a singular limit of <u>hyperelliptic</u> curves. This occurs if $b_k = -c_k$ for all k. In fact, when this is satisfied, if t is the coordinate on $\mathbb{P}^1$, let

$$x = t^2$$

$$y = t\cdot\prod_{i=1}^{g}(t^2-b_i^2).$$

Then $x(b_k) = x(c_k)$, $y(b_k) = y(c_k)$, and the 2 functions x,y embed the singular curve $C-\{\infty\}$ into $\mathbb{C}^2$. The image is defined by $y^2 = x \cdot \Pi(x-b_i^2)^2$, which is a limit of equations $y^2 = f_{2g+1}(x)$ for smooth hyperelliptic curves of genus g.

Recall from the beginning of this section that the invariant vector field on Jac C associated to a point $x_0 \in C$ is

$$D_{x_0} = \sum_i \frac{b_i - c_i}{(b_i-x_0)(c_i-x_0)} \lambda_i \frac{\partial}{\partial\lambda_i} .$$

If $b_k = -c_k$, then

$$D_{x_0} = 2\cdot \sum_i \frac{b_i\cdot x_0^{-2}}{1-b_i^2 x_0^{-2}} \lambda_i \frac{\partial}{\partial\lambda_i}$$

$$= 2\cdot (x_0^{-2} \sum_i b_i\lambda_i \frac{\partial}{\partial\lambda_i} + x_0^{-4} \sum_i b_i^3\lambda_i \frac{\partial}{\partial\lambda_i} +\cdots).$$

The vector field associated to the point at infinity is therefore:

$$D_\infty = \sum_i b_i\lambda_i \frac{\partial}{\partial\lambda_i} ,$$

and the singular $\wp$-function is:

$$D_\infty^2\log \tau_C(\lambda_1,\ldots,\lambda_g) = (\sum b_i\lambda_i \frac{\partial}{\partial\lambda_i})^2\log \sum_{S\subset 1..g} (-1)^{\#S} \cdot \prod_{\substack{i<j\\ i,j\in S}} \left(\frac{b_i-b_j}{b_i+b_j}\right)^2 \cdot \prod_{i\in S}\left(\lambda_i\cdot \prod_{j\neq i}\frac{b_j+b_i}{b_j-b_i}\right)$$

To obtain a solution to KdV, we need merely substitute

$$\lambda_i = e^{(e_i+b_ix-2b_i^3t)} .$$

for any $e_1,\ldots,e_g$; or absorbing the factor $-\prod_{j\neq i}\frac{b_j+b_i}{b_j-b_i}$ in the e_i,

$$f(x,t) = \left(\frac{\partial}{\partial x}\right)^2 \log \sum_{S\subset\{1,\ldots,g\}} \prod_{\substack{i<j\\ i,j\in S}} \left(\frac{b_i-b_j}{b_i+b_j}\right)^2 \cdot \prod_{i\in S} e^{(e_i+b_ix-2b_i^3t)} .$$

These are precisely the g-soliton solutions of KdV.

The famous asymptotic properties of g-solitons (that for $t \ll 0$, it splits up into g widely separated blobs, which interact for moderate values of t, and which for $t \gg 0$ split up again into the same g blobs, with the same shape but with a phase shift) can all be deduced very simply from the above formula and the fact that $D_\infty^2 \log \tau_C$ extends to a continuous function on the compactification $\overline{\text{Jac } C}$ described above of the generalized jacobian. To get a real-valued function f(x,t), assume that all b_i are real, and define

$$\sigma\colon \mathbb{R}^2 \longrightarrow (\mathbb{C}^*)^g \cong \text{Jac } C \subset \overline{\text{Jac } C}$$

by

$$\sigma(x,t) = \left(\cdots, -\prod_{j\neq i}\frac{b_j-b_i}{b_j+b_i}\, e^{e_i+b_ix-2b_i^3t}, \cdots\right).$$

Then

$$f(x,t) = (D_\infty^2 \log \tau_C)(\sigma(x,t)).$$

As shown above, $D_\infty^2 \log \tau_C$ extends to a continuous function on $\overline{\text{Jac } C}$. In fact, it is zero at the "most singular" point P_∞ given by letting all coordinates λ_i on $(\mathbb{C}^*)^g$ tend to 0 or ∞. To see this, write

$$\tau_C = \sum_{S\subset\{1,\cdots,g\}} a_S\lambda^S .$$

Then if

$$b_S = \sum_{i\in S} b_i ,$$

$$D_\infty^2 \log \tau_C = \frac{\Sigma a_S\lambda^S \cdot \Sigma a_S b_S^2 \lambda^S - (\Sigma a_S b_S \lambda^S)^2}{(\Sigma a_S\lambda^S)^2} .$$

Note that all terms λ^{2S} in the numerator cancel out while for every S, the denominator has a λ^{2S} term since $a_S \neq 0$. Thus

$$D_\infty^2 \log \tau_C(P_\infty) = 0.$$

Therefore, for all $\varepsilon > 0$, there is a neighborhood U_ε of P_∞ in $\overline{\text{Jac } C}$ such that:

$$P \in U_\varepsilon \Longrightarrow |D_\infty^2 \log \tau_C(P)| < \varepsilon.$$

Therefore, there is a constant c such that if

$$|x-2b_i^2 t| > C, \quad \text{all } 1 \le i \le g \Longrightarrow \sigma(x,t) \in U_\varepsilon$$
$$\Longrightarrow |f(x,t)| < \varepsilon.$$

Thus the effective support of f(x,t) is a set of g bands

$$|x-2b_i^2 t| \le C$$

representing "blobs" moving with distinct positive velocities $2b_i^2$.

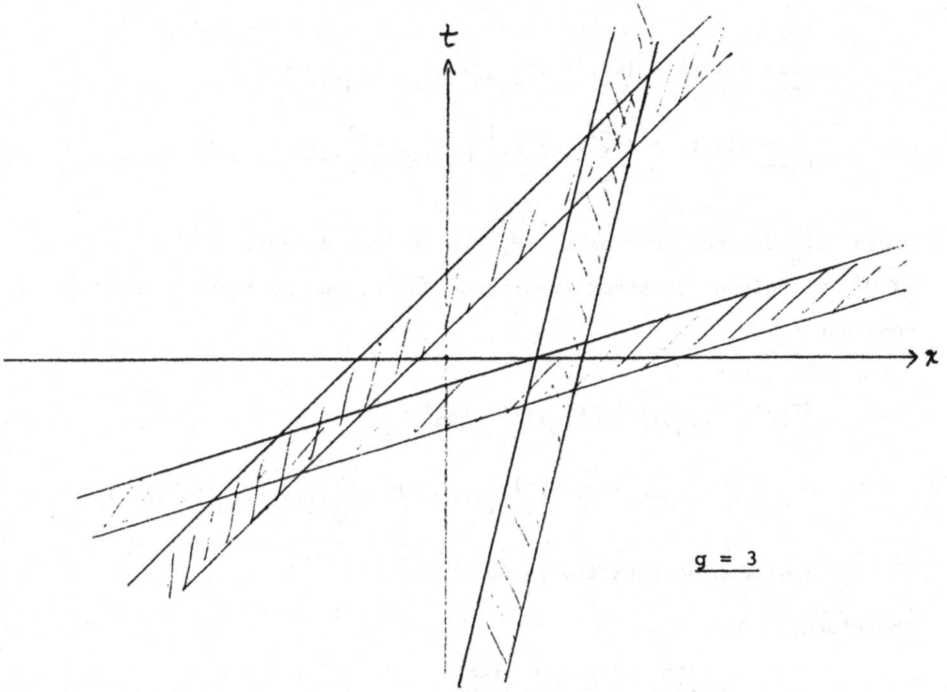

g = 3

Moreover, if $t \longrightarrow \pm\infty$ and we stay in the i_o^{th} band, then

$$|x-2b_i^2 t| \longrightarrow \infty, \quad i \neq i_0$$

and lim $\sigma(x,t)$ will lie on the i_o^{th} 1-dimensional strata

$$J_{i_0} = \{(\lambda_1,\cdots,\lambda_g)\,|\,\lambda_{i_0} \in \mathbb{C}^*, \text{ but } \lambda_i \in \{0,\infty\} \text{ if } i \neq i_0\}.$$

In fact, fix the value $z = x-2b_{i_0}^2 t$ and let $t \longrightarrow \pm\infty$. Then for some choice of $\varepsilon_i \in \{0,\infty\}$, $(i \neq i_0)$,

$$\lim_{t\to-\infty} \sigma(x,t) = (\varepsilon_1,\cdots,\varepsilon_{i_0-1}, \lambda_{i_0}, \varepsilon_{i_0+1},\cdots,\varepsilon_g)$$

$$\lim_{t\to+\infty} \sigma(x,t) = (\varepsilon_1^{-1},\cdots,\varepsilon_{i_0-1}^{-1}, \lambda_{i_0}, \varepsilon_{i_0+1}^{-1},\cdots,\varepsilon_g^{-1})$$

where λ_{i_0} depends only on z. By the theorem describing how $(\mathbb{P}^1)^g$ is "glued" together to produce $\overline{\text{Jac } C}$, we see that for some constant η_{i_0},

$$(\varepsilon_1,\cdots,\varepsilon_{i_0-1}, \lambda, \varepsilon_{i_0+1},\cdots,\varepsilon_g)$$

$$\sim (\varepsilon_1^{-1},\cdots,\varepsilon_{i_0-1}^{-1}, e^{\eta_{i_0}}\lambda, \varepsilon_{i_0+1}^{-1},\cdots,\varepsilon_g^{-1}), \text{ all } \lambda$$

($\sim$ meaning equality in $\overline{\text{Jac } C}$).

Therefore,

$$\lim_{\substack{t\to-\infty \\ x-2b_{i_0}t=z}} f(x,t) = \lim_{\substack{t\to+\infty \\ x-2b_{i_0}t=z+(\eta_{i_0}/b_{i_0})}} f(x,t) ,$$

i.e., for $t \to -\infty$ or $t \to \infty$, $f(x,t)$ has the same shape in each band except for a phase shift. The fact that this shape is a single "wave" moreover is more or less a consequence of the simple fact that on each 1-dimensional stratum J_{i_0}, the rational function $\tau_C(\lambda)$ tends asymptotically to $(1+\lambda)\lambda^s$ (in a suitable coordinate $\lambda \in \mathbb{C}^*$), i.e., up to the scale factor λ^s, has a single negative zero. When you set $\lambda = e^{bx}$ and take logarithmic derivatives, f will have a single pair of complex conjugate poles closest to the real axis and these give its wave shape. More generally, the zeroes of τ_C on $\overline{\text{Jac}(C)}$ give poles of $f(x,t)$ but only for complex values of x,t and f will have large values along the real points near these poles.

Resolution of algebraic equations by theta constants

Hiroshi UMEMURA

The history of algebraic equations is very long. The necessity and the trial of solving algebraic equations existed already in the ancient civilizations. The Babylonians solved equations of degree 2 around 2000 B.C. as well as the Indians and the Chinese. In the 16th century, the Italians discovered the resolutions of the equations of degree 3 and 4 by radicals known as Cardano's formula and Ferrari's formula. However in 1826, Abel [1] (independently about the same epoch Galois [7]) proved the impossibility of solving general equations of degree ≥ 5 by radicals. This is one of the most remarkable event in the history of algebraic equations. Was there nothing to do in this branch of mathematics after the work of Abel and Galois? Yes, in 1858 Hermite [8] and Kronecker [15] proved that we can solve the algebraic equation of degree 5 by using an elliptic modular function. Since $\sqrt[n]{a} = \exp((1/n)\log a)$ which is also written as $\exp((1/n)\int_1^a (1/x)dx)$, to allow only the extractions of radicals is to use only the exponential. Hence under this restriction, as we learn in the Galois theory, we can construct only compositions of cyclic extensions, namely solvable extentions. The idea of Hermite and Kronecker is as follows; if we use another transcendental function than the exponential, we can solve the algebraic equation of degree 5. In fact their result is analogous to the formula $\sqrt[n]{a} = \exp(1/n)\int_1^a (1/x)dx)$. In the quintic equation they replace the exponential by an elliptic modular function and the integral $\int (1/x)dx$ by elliptic integrals. Kronecker [15] thought the resolution of the equation of degree 5 by an elliptic modular function would be a special case of a more general theorem which might exist. Kronecker's idea was realized in few cases by Klein [11], [13]. Jordan [10] showed that we can solve any algebraic equation of higher degree by modular functions. Jordan's idea is clarified by Thomae's formula, §8 Chap. III (cf. Lindemann [16]). In this appendix, we show how we can deduce from Thomae's formula the resolution of algebraic equations by a Siegel modular function which is explicitely expressed by theta constants (Theorem 2). Therefore Kronecker's idea is completely realized. Our resolution of higher algebraic equations is also similar to the formula $\sqrt[n]{a} = \exp((1/n)\int_1^a (1/x)dx)$. In our resolution the exponential is replaced by the Siegel modular function and the integral $\int (1/x)dx$ is replaced by hyperelliptic integrals. The existance of such resolution shows that the theta function is useful not only for non-linear differential equations but also for algebraic equations.

Let us fix some notations. We follow in principle the covention of Chap. III. Let $F(X)$ be a polynomial of odd degree $2g+1$ with coefficients in the complex number field $\mathbb{C}$. We assume that the equation $F(X) = 0$ has only simple roots so that $Y^2 = F(X)$ defines a hyperelliptic curve C of genus g. Then C is a two sheeted covering of $\mathbb{P}^1$ ramified at the roots of $F(X) = 0$ and at ∞. Let $x_1, x_2, \cdots, x_{2g+1}$ be the roots of $F(X) = 0$. Let us set $B' = \{1,2,\cdots,2g+1\}$. For two subsets S, T of B', we put $S \circ T = S \cup T - S \cap T$. η_{2i-1} is defined as the $2\times g$ matrix

$$\begin{bmatrix} 0\cdots 0 & \overset{i^{th}\text{ place}}{\frac{1}{2}} & 0\cdots 0 \\ \frac{1}{2}\cdots\frac{1}{2} & 0 & \cdots\cdots 0 \end{bmatrix}$$

and η_{2i} is the $2\times g$ matrix

$$\begin{bmatrix} 0\cdots 0 & \overset{i^{th}\text{ place}}{\frac{1}{2}} & 0\cdots 0 \\ \frac{1}{2}\cdots\cdots & \frac{1}{2} & 0\cdots 0 \end{bmatrix}.$$

For all $T \subset B$, the sum $\sum_{k\in T} \eta_k$ is denoted by η_T. Classically the period matrix Ω of C is calculated with respect to the normalized basis of $H_1(C, \mathbb{Z})$ in §5, Chap. III. Thus Ω is determined when we fix not only $F(x)$ but also the order of its roots. Finally we put $U = \{1,3,\cdots,2g+1\}$, the subset of B' consisting of all the odd numbers of B'. For row vectors $m_1, m_2 \in \mathbb{R}^g$, $z \in \mathbb{C}^g$ and a symmetric $g\times g$ matrix τ with positive definite imaginary part, we define the theta function

$$\theta\begin{bmatrix} m_1 \\ m_2 \end{bmatrix}(z,\tau) = \sum_{\xi\in\mathbb{Z}^g} e\left(\tfrac{1}{2}(\xi+m_1)\tau\,{}^t(\xi+m_1) + (\xi+m_1)\,{}^t(Z+m_2)\right)$$

where $e(x) = \exp(2\pi i x)$. The theta constant $\theta\begin{bmatrix} m_1 \\ m_2 \end{bmatrix}(0,\tau)$ will be denoted by $\theta\begin{bmatrix} m_1 \\ m_2 \end{bmatrix}(\tau)$ for short.

Theorem 1. The following equality holds;

$$\frac{x_1-x_3}{x_1-x_2}$$

$$= \left(\theta\begin{bmatrix} \frac{1}{2} & 0\cdots 0 \\ 0 & \cdots & 0 \end{bmatrix}(\Omega)^4 \theta\begin{bmatrix} \frac{1}{2} & \frac{1}{2} & 0\cdots 0 \\ 0 & \cdots\cdots & 0 \end{bmatrix}(\Omega)^4 + \theta\begin{bmatrix} 0\cdots 0 \\ 0\cdots 0 \end{bmatrix}(\Omega)^4 \theta\begin{bmatrix} 0 & \frac{1}{2} & 0\cdots 0 \\ 0 & \cdots\cdots & 0 \end{bmatrix}(\Omega)^4\right.$$

$$\left. - \theta\begin{bmatrix} 0 & \cdots & 0 \\ \frac{1}{2} & 0\cdots 0 \end{bmatrix}(\Omega)^4 \theta\begin{bmatrix} 0 & \frac{1}{2} & 0\cdots 0 \\ \frac{1}{2} & 0 & \cdots & 0 \end{bmatrix}(\Omega)^4\right) \Big/ \left(2\theta\begin{bmatrix} \frac{1}{2} & 0\cdots 0 \\ 0 & \cdots & 0 \end{bmatrix}(\Omega)^4 \theta\begin{bmatrix} \frac{1}{2} & \frac{1}{2} & 0\cdots 0 \\ 0 & \cdots\cdots & 0 \end{bmatrix}(\Omega)^4\right).$$

The theorem is deduced from Theorem 8.1, §8, Chap. III, by carrying out precisely the calculation indicated in the proof of Corollary 8.13 and form the formula $\theta[{m_1+\xi_1 \atop m_2+\xi_2}](z,\tau) = e(m_1^t\xi_2)\cdot \theta[{m_1 \atop m_2}](z,\tau)$ for $\xi_1, \xi_2 \in \mathbb{Z}^g$ (see for example Igusa [9], Chap. I §10, $(\theta,2)$ p.49). In fact for a division $B = V_1 \amalg V_2 \amalg \{k\}$ with $\#V_1 = \#V_2 = g$, it follows from Theorem 8.1, §8, Chap. III,

$$(1.1)\quad \theta[\eta_{(V_2+k)\circ U}](\Omega)^4 = c(-1)^{\#(U-(V_2+k))} \prod_{i\in V_2+k,\, j\in V_1} (x_i-x_j)^{-1}$$

because $(V_2+k\circ U)\circ U = V_2+k$ here the union $V_2\amalg\{k\}$ is denoted by V_2+k. Theorem 8.1, §8, Chap. III for $S = (V_1+k)\circ U$ gives

$$(1.2)\quad \theta[\eta_{(V_1+k)\circ U}](\Omega)^4 = c(-1)^{\#(U-(V_1+k))} \prod_{i\in V_1+k,\, j\in V_2} (x_i-x_j)^{-1}.$$

Dividing (1.1) by (1.2), we get

$$(1.3)\quad \frac{\theta[\eta_{(V_2+k)\circ U}](\Omega)^4}{\theta[\eta_{(V_1+k)\circ U}](\Omega)^4}$$

$$= (-1)^{\#(U-(V_2+k))+\#(U-V_1+k))} \frac{\prod_{\ell\in V_1+k,\, m\in V_2}(x_\ell-x_m)}{\prod_{i\in V_2+k,\, j\in V_1}(x_i-x_j)}$$

$$= (-1)^{k+1} \frac{\prod_{i\in V_2}(x_k-x_i)}{\prod_{i\in V_1}(x_k-x_i)}.$$

Let us consider a division $B' = \{1,2,3\}\amalg\{2n \mid 2\le n\le g\}\amalg\{2n+1 \mid 2\le n\le g\}$. Putting $V_3 = \{2n \mid 2\le n\le g\}$, $V_4 = \{2n+1 \mid 2\le n\le g\}$, we apply (1.3) for $k=1$, $V_1 = V_3+2$, $V_2 = V_4+3$;

$$(1.4)\quad \frac{\theta[\eta_{(V_4+3+1)\circ U}](\Omega)^4}{\theta[\eta_{(V_3+2+1)\circ U}](\Omega)^4} = \frac{\prod_{i\in V_4+3}(x_1-x_i)}{\prod_{i\in V_3+2}(x_1-x_i)}.$$

Next (1.3) for $k=1$, $V_1=V_4+2$, $V_2=V_3+3$ is

$$(1.5)\qquad \frac{\theta[\eta_{(V_3+3+1)\circ U}](\Omega)^4}{\theta[\eta_{(V_4+2+1)\circ U}](\Omega)^4} = \frac{\prod_{i\in V_3+3}(x_1-x_i)}{\prod_{i\in V_4+2}(x_1-x_i)}.$$

Multiplying (1.4) with (1.5), we get

$$(1.6)\qquad \frac{\theta[\eta_{(V_4+3+1)\circ U}](\Omega)^4\theta[\eta_{(V_3+3+1)}](\Omega)^4}{\theta[\eta_{(V_3+2+1)\circ U}](\Omega)^4\theta[\eta_{(V_4+2+1)}](\Omega)^4} = \frac{(x_1-x_3)^2}{(x_1-x_2)^2}.$$

For the above division $B' = \{1,2,3\} \amalg \{2n \mid 2\le n\le g\} \amalg \{2n+1 \mid 2\le n\le g\}$ if we interchange 1 and 2, then (1.6) becomes

$$(1.7)\qquad \frac{\theta[\eta_{(V_4+3+2)\circ U}](\Omega)^4\theta[\eta_{(V_3+3+2)\circ U}](\Omega)^4}{\theta[\eta_{(V_3+1+2)\circ U}](\Omega)^4\theta[\eta_{(V_4+1+2)\circ U}](\Omega)^4} = \frac{(x_2-x_3)^2}{(x_2-x_1)^2}.$$

We notice the following identity,

$$(1.8)\qquad \frac{x_1-x_3}{x_1-x_2} = \frac{1}{2}\{1 + (\frac{x_1-x_3}{x_1-x_2})^2 - (\frac{x_2-x_3}{x_2-x_1})^2\}.$$

It follows from (1.6),(1.7) and (1.8)

$$(1.9)\qquad \frac{x_1-x_3}{x_1-x_2} = (\theta[\eta_{(V_3+2+1)\circ U}](\Omega)^4\theta[\eta_{(V_4+2+1)\circ U}](\Omega)^4$$

$$+ \theta[\eta_{(V_4+3+1)\circ U}](\Omega)^4\theta[\eta_{(V_3+3+1)\circ U}](\Omega)^4$$

$$- \theta[\eta_{(V_4+3+2)\circ U}](\Omega)^4\theta[\eta_{(V_3+3+2)\circ U}](\Omega)^4)/$$

$$(2\theta[\eta_{(V_3+2+1)\circ U}](\Omega)^4\theta[\eta_{(V_4+2+1)\circ U}(\Omega)^4).$$

The theta characteristics in (1.9) are half integral. Theorem now follows from the following formula: for ξ_1,ξ_2 in $\mathbb{Z}^g$,

$$\theta\begin{bmatrix} m_1+\xi_1 \\ m_2+\xi_2 \end{bmatrix}(z,\tau) = e(m_1^t\xi_2)\theta\begin{bmatrix} m_1 \\ m_2 \end{bmatrix}(z,\tau).$$

We notice that by the transformation formula, the right hand side of the equality in Theorem 1 is a Siegel modular function of level 2 (see Igusa [9], Chap. 5 §1, Corollary).

A marvellous application of Theorem 1 is the resolution of the algebraic equation by a Siegel modular function.

Theorem 2. Let

(2.1) $$a_0X^n + a_1X^{n-1} + \cdots + a_n = 0, \quad a_0 \neq 0, \ a_i \in \mathbb{C} \quad (0 \le i \le n)$$

be an algebraic equation irreducible over a certain subfield of $\mathbb{C}$, then a root of the algebraic equation (2.1) is given by

(2.2) $$\Big(\theta\begin{bmatrix} \frac{1}{2} & 0\cdots 0 \\ 0 & \cdots\ 0 \end{bmatrix}(\Omega)^4\theta\begin{bmatrix} \frac{1}{2}\ \frac{1}{2} & 0\cdots 0 \\ 0 & \cdots\cdots\ 0 \end{bmatrix}(\Omega)^4 + \theta\begin{bmatrix} 0\cdots 0 \\ 0\cdots 0 \end{bmatrix}(\Omega)^4\theta\begin{bmatrix} 0\ \frac{1}{2} & 0\cdots 0 \\ 0 & \cdots\cdots\ 0 \end{bmatrix}(\Omega)^4$$
$$- \theta\begin{bmatrix} 0 & 0\cdots 0 \\ \frac{1}{2} & 0\cdots 0 \end{bmatrix}(\Omega)^4\theta\begin{bmatrix} 0 & \frac{1}{2} & 0\cdots 0 \\ \frac{1}{2} & 0 & \cdots\ 0 \end{bmatrix}(\Omega)^4\Big)\Big/\Big(2\theta\begin{bmatrix} \frac{1}{2} & 0\cdots 0 \\ 0 & \cdots\ 0 \end{bmatrix}(\Omega)^4\theta\begin{bmatrix} \frac{1}{2}\ \frac{1}{2} & 0\cdots 0 \\ 0 & \cdots\cdots\ 0 \end{bmatrix}(\Omega)^4\Big),$$

where Ω is the period matrix of a hyperelliptic curve $C : Y^2 = F(X)$ with $F(X) = X(X-1)(a_0X^n + a_1X^{n-1} + \cdots + a_n)$ for n odd and $F(X) = X(X-1)(X-2)(a_0X^n + a_1X^{n-1} + \cdots + a_n)$ for n even. More precisely let $\alpha_1, \alpha_2, \cdots, \alpha_n$ be the roots of equation (2.1). Then Ω is the period matrix of the hyperelliptic curve C with respect to the classical normalized basis of $H_1(C, \mathbb{Z})$ when the roots of $F(X) = 0$ are ordered as follows : for n odd $x_1 = 0$, $x_2 = 1$, $x_{i+2} = \alpha_i$ $(1 \le i \le n)$ and for n even $x_1 = 0$, $x_2 = 1$, $x_{i+2} = \alpha_i$ $(1 \le i \le n)$, $x_{n+3} = 2$. The root α_1 of equation (2.1) is given by (2.2).

Proof. It follows from the assumption that the equation is irreducible over a subfield of $\mathbb{C}$, $F(X) = 0$ has only simple roots. Since $(x_1 - x_3)/(x_1 - x_2) = x_3 = \alpha_1$, Theorem 2 follows from Theorem 1.

To determine the period Ω we have to number the roots of the algebraic equation. Even if we don't know the precise roots of the equation, the numbering can be done once we can separate the roots of the algebraic equation. The complex Sturm theorem says that there exists an algorithm of separating the roots of the algebraic equation (Weber [19], I §103, §104). Therefore Theorem 2 is a resolution of an algebraic equation by a Siegel modular function. Compared with the formula $\sqrt[n]{a} = \exp((1/n)\log a)$ $= \exp((1/n)\int_1^a (1/x)dx)$, in our theorem the exponential is replaced by the Siegel modular function (2.2) and the integral $\int_1^a (1/x)dx$ is replaced by hyperelliptic integrals $\int (x^i/\sqrt{F(x)})dx$, $0 \le i \le g-1$ which determine the period Ω.

Let us compare our Theorem with the result due Hermite [8], Kronecker [15] and Klein [12] on the resolution of the quintic algebraic equation by an elliptic modular function. Their theory sticks to the modular variety of elliptic curves with level five structure (cf. Chap. I). Let H be the upper half plane and Γ_n be the principal congruence subgroup of level n $\{\begin{pmatrix} a & b \\ c & d \end{pmatrix} \in SL_2(\mathbb{Z}) \mid b \equiv c \equiv 0,\ a \equiv d \equiv 1 \bmod n\}$. Γ_n operates on H in usual way and the quotient variety H/Γ_n is the modular variety of elliptic curves with level n structure. The function field $\mathbb{C}(H/\Gamma_n)$ has a model $\mathbb{Q}(H/\Gamma_n)$ over $\mathbb{Q}$ and the morphism $\pi: H/\Gamma_n \to H/\Gamma_1$ descends giving an inclusion $\mathbb{Q}(H/\Gamma_1) \hookrightarrow \mathbb{Q}(H/\Gamma_n)$ (see Deligne et Rapaport [4]). The natural projection $H/\Gamma_n \to H/\Gamma_1$ is a Galois covering with group $\Gamma_1/\pm\Gamma_n$. Therefore $H/\Gamma_5 \to H/\Gamma_1$ is a Galois covering with group $\Gamma_1/\pm\Gamma_5$ which is isomorphic to the alternating group $\mathfrak{A}_5$ of degree 5. Since H/Γ_1 is a rational curve and its coordinate ring $\mathbb{Q}[H/\Gamma_1]$ is a polynomial ring $\mathbb{Q}[j(\omega)]$, $\mathbb{Q}(H/\Gamma_5)/\mathbb{Q}(H/\Gamma_1)$ is a one parameter family of Galois extensions with group $\mathfrak{A}_5$. The key point is this family contains any Galois extension with group $\mathfrak{A}_5$ in $\mathbb{C}$. To be more precise, since $\mathfrak{A}_5$ has a subgroup of index 5, there exists an extension (resolvent) $\mathbb{Q}(H/\Gamma_5) \supset F \supset \mathbb{Q}(H/\Gamma_1)$ with $[F, \mathbb{Q}(H/\Gamma_1)] = 5$. Moreover one can show among such resolvents there is a particular one described explicitely by using the Dedekind η function : There exists a resolvent of degree 5 of $Q(H/\Gamma_5)/\mathbb{Q}(H/\Gamma_1)$ given by an equation

$$(2.3) \qquad w^5 + b_1 w^4 + b_2 w^3 + b_3 w^3 + b_4 w + b_5 = j(\omega), \; b_i \in \mathbb{Q} \; (1 \le i \le 5)$$

and the solutions $w_i(\omega)$ $(1 \le i \le 5)$ of equation (2.3) are explicitely written by the Dedekind η function. Now given a general quintic equation over a subfield k of $\mathbb{C}$

$$(2.4) \qquad X^5 + a_1 X^4 + a_2 X^3 + a_3 X^2 + a_4 X + a_5 = 0, \quad a_i \in k, \; (1 \le i \le 5).$$

Then it is easy to see that by a Tschirnhausen transformation involving only the extractions of square and cube roots, the resolution of the given equation (2.4) is reduced to that of

$$(2.5) \qquad X^5 + b_1 X^4 + b_2 X^3 + b_3 X^2 + b_4 X + a_5' = 0$$

where a_5' is in a solvable extension of $k(a_i)_{1 \le i \le 5}$ obtained by adjunction of square and cube roots (Weber [19], I §60, §80,

§81). Next we look for a point $\omega_0 \in H$ such that $a_5' = b_5 - j(\omega_0)$. This procedure depends on elliptic integrals. Recall for an elliptic curve $C : y^2 = 4x^3 - g_2x - g_3$ the modular invariant j of C is equal to $2^6 \cdot 3^2 \cdot g_2^3/(g_2^3 - 27g_3^2)$. We solve in $\mathbb{C}$ $b_5 - a_5' = 2^6 \cdot 3^2 \cdot a^3/(a^3 - 27b^2)$ for unknowns a, b. This is done by extractions of a square or cube root. Then the period ω_0 of the elliptic curve $C : y^2 = 4x^3 - ax - b$ is calculated by elliptic integrals $\int_\gamma 1/\sqrt{4x^3-ax-b}\, dx$ for suitable paths γ and $j(\omega_0) = b_5 - a_5'$. Therefore $w_i(\omega_0)$ $(1 \le i \le 5)$ are the solutions of the equation (2.5) hence the given equation (2.4) is solved. If we try to solve a quintic equation by Theorem 2, it is simpler than the above mentioned classical method because in our theory the Tschirnhausen transformation is not involved. But we need a modular function of genus 3.

Remark 3. Let

$$f(X) = a_0X^n + a_1X^{n-1} + \cdots + a_n = 0 \quad a_0 \neq 0,\ a_i \in \mathbb{C}\ (0 \le i \le n) \tag{3.1}$$

be a general algebraic equation of even degree $n = 2g+2 \ge 4$ over a subfield k of $\mathbb{C}$. We do not want to clarify the word "general". Then considering $f(X)$ itself as $F(X)$ instead of multiplying X, $(X-1)$ or $(X-2)$, we can show that for $f(X) = F(X)$, the values of the modular function in Theorem 1 for all the orders of the roots of $F(X) = 0$, generate the Galois extension of (3.1) over k. In this form, the back ground of our theorem is clear. Let $\mathcal{H}_g^{(2)}$ be the moduli space of $(C, (x_1, x_2, \cdots, x_{2g+2}))$, C a hyperelliptic curve of genus g and $(x_1, x_2, \cdots, x_{2g+2})$, the (ordered) set of the Weierstrass points as in §8

Chap. III. The symmetric group $\mathfrak{S}_{2g+2}$ operates on $\mathcal{H}_g^{(2)}$ as permutations of the Weierstrass points. By Chap. III §2, Lemma 2.4 and §6, Proposition 6.1, $\mathcal{H}_g^{(2)}$ is a subvariety of the modular variety M_2 of the principally polarized abelian varieties of dimension g with level 2 structure. Let M_1 be the modular variety of the principally polarized abelian varieties of dimension g. Then there is a canonical morphism $M_2 \to M_1$ of forgetting the level 2 structure. This morphism is a Galois covering with group $Sp_{2g}(\mathbb{Z}/2\mathbb{Z})$. The Galois group of (3.1) which is a subgroup of $\mathfrak{S}_{2g+2}$, interchanges the Weierstrass points of the hyperelliptic curve $C : Y^2 = F(X)$. This operation of $\mathfrak{S}_{2g+2}$ on the Weierstrass points induces a faithful representation $\mathfrak{S}_{2g+2} \to Sp(J(C)_2) = Sp_{2g}(\mathbb{Z}/2\mathbb{Z})$ by Chap. III §6, Proposition 6.3. Therefore the equation (3.1) is solved in a specialization of the Galois covering $M_2 \to M_1$. The specialization involves the modular function in Theorem 1 and the hyperelliptic integrals.

Remark 4. Finally we notice that Theorem 2 is similar to Jacobi's formula : Setting $K = \int_0^1 dx/\sqrt{(1-x^2)(1-k^2x^2)}$, $iK' = \int_1^{1/k} dx/\sqrt{(1-x^2)(1-k^2x^2)}$ and $\omega = iK'/K$, we have $k = \theta_{10}^2(0,\omega)/\theta_{00}^2(0,\omega)$. Jacobi's formula solves a quadratic equation $1-k^2x^2 = 0$ by theta constants and elliptic integrals.

Bibliography

[1] Abel, N., H., Beweis der Unmöglichkeit algebraische Gleichugen von höheren Graden als dem vierten allgemein aufzulösen, J. für die reine und angew. Math., Bd. 1 (1826) 65-84.

[2] Belardinelli, G., Fonctions hypergéométriques de plusieurs variables et résolution analytique des équations algebriques générales, Memorial des sc. math., Gauthier-Villars, Paris (1960).

[3] Coble, A., B., The equation of the eighth degree, Bull. Amer. Math. Soc., 30 (1924), 301-313.

[4] Deligne, P. et Rapaport, M., Les schemas de modules de courbes elliptiques, Modular functions of one variable II, Lecture Notes in Math., 349, Springer-Verlag, Berlin, Heidelberg, New York (1973).

[5] Enriques, F., Sur les problèmes qui se rapportent à la résolution des équations algébriques renfermant plusieurs inconnues, Math. Ann., Bd. 51 (1899), 134-153.

[6] Fricke, R., Lehrbuch der Algebra, Friedr. Vieweg & Sohn, Braunschweig (1924).

[7] Galois, E., Ecrits et mémoires mathématiques, Gauthier-Villars, Paris (1962).

[8] Hermite, Ch., Sur la resolution de l'equation du cinquième degré, C. R. Acad. Sc., t. 46 (1858) 508-515.

[9] Igusa, J., Theta functions, Springer-Verlag, Berlin, Heidelberg, New York (1972).

[10] Jordan, C., Traité des substitutions et des équations algébriques, Gauthier-Villars, Paris (1870).

[11] Klein, F., Gleichungen vom siebenten und achten Grade, Math. Annalen, Bd. 15 (1879) 251-282.

[12] ————, Vorlesungen über das Icosaeder und die Auflösung der Gleichungen vom fünften Grade, Teubner, Leipzig (1884).

[13] ————, Sur la resolution, par les fonctions hyper-elliptiques, de l'equation du vingt-septième degré de laquelle dépend la determination des vingt-sept droites d'une surface cubique, Atti Rend. R. Acad. dei Lincei, Ser. 5a, vol. 8 (1899).

[14] ————, Lectures on Mathematics, Evanston Colloquium, Macmillan, New York (1894).

[15] Kronecker, L., Sur la résolution de l'equation du cinquième degré, C. R. Acad. Sc., t. 46 (1858) 1150-1152.

[16] Lindemann, F., Ueber die Auflösung algebraischer Gleichungen durch transcendente Functionen I, II, Göttingen Nach. (1884) 245-248, (1892) 292-298.

[17] Mellin, Hj., Résolution de l'équation algébrique générale à l'aide de la fonction Γ, C. R. Acad. Sc., t. 172 (1921), 658-661.

[18] Thomae, J., Beitrag zur Bestimmung von $\theta(0,0,\cdots,0)$ durch die Klassenmoduln algebraischer Functionen, J. für die reine und angew. Math., Bd. 71 (1870) 201-222.

[19] Weber, H., Lehrbuch der Algebra, Reprint, Chelsea New York.